MATHEMATICAL PHYSICS

MATHEMATICAL PHYSICS

Jyoti Kumar

RAJAT PUBLICATIONS
New Delhi

Mathematical Physics
by *Jyoti Kumar*

This edition is published by Rajat Publications
4740/23, Ansari Road Daryaganj,
New Delhi - 110002

ISBN: 978-81-7880-380-7

Edition 2026

Printed and bound in India

RAJAT
PUBLICATIONS

4740/23, Ansari Road, Daryaganj,
New Delhi- 110002 (India)
Tel:- +91-11-23267924
Mob:- +91-9811484973, +91-9811105389
E-mail: rajat_publication@yahoo.com

Preface

Mathematical Physics has been emerging as a new branch of Mathematical Sciences, rapidly developing and gaining popularity in the last few years. It is closely related to Idempotent Mathematics. The authors have endeavoured to inculcate the latest developments and study on various research papers on the subject.

The major topics on Mathematical Physics, including: Physical Dimensions and Fundamental Units, Mechanics and Dynamics, Waves and Vibrations, Classical Electromagnetic Theory, Special Relativity, etc., have been comprehensively and lucidily dealt with in this book.

We feel pleasure in bringing out this book and hope that it will prove useful for the students appearing for different graduate, honours, competitive examinations and postgraduation of various Indian and Foreign Universities.

I am grateful to the publishers for all cooperation and bringing out the book in time.

Jyoti Kumar

Contents

1

Physical Dimensions and Fundamental Units

FUNDAMENTAL UNIT

A set of fundamental units is a set of units for physical quantities from which every other unit can be generated.

In the language of measurement, *quantities* are quantifiable aspects of the world, such as time, distance, velocity, mass, momentum, energy, and weight, and *units* are used to describe their measure. Quantities are sometimes also referred to as *dimensions* (i.e. "The dimensions of the package to be mailed are 12" × 8" × 4".") although the term is normally meant to apply to the class of physical quantity (length, time, mass, momentum, energy, etc.) being measured (see dimensional analysis).

A system of fundamental quantities (or sometimes fundamental dimensions) is such that every other physical quantity (or dimension of physical quantity) can be generated from them. The traditional fundamental dimensions of physical quantity are mass, length, time, and temperature, but in principle, other fundamental quantities could be used. Some physicists have not recognized temperature as a fundamental dimension of physical quantity since it simply expresses the energy per particle per degree of freedom which can be expressed in terms of energy (or mass, length, and time). In addition, some physicists recognize electric charge as a separate fundamental dimension of physical quantity, even if it has been expressed in terms of mass, length, and time in unit systems such as the electrostatic cgs system. There are also physicists who have cast doubt on the very existence of incompatible fundamental quantities

The fundamental units of the SI system are kilogram, second, ampere, kelvin, and mole. The metre and candela were formerly fundamental units but have been redefined in terms of the other units. They are still considered SI base units, however, mainly for historical reasons. Other units such as the newton, joule, and volt can be derived from the SI base units and are therefore termed SI derived units.

The dimension speed, for example, is length divided by time, so its unit m/s can be generated from the fundamental unit of seconds (the metre being also defined in terms of the second).

An important basic fact of dimensional analysis is that the fundamental units can be regarded as the basis of a special kind of vector space, the space of all units. This is a vector space over the field of rational numbers where the vector addition is given by the multiplication of units and the scalar multiplication is exponentiation of units.

Not all physically salient values have units: dimensionless numbers occur in many fields of science. In fact, dimensionless quantities are fundamentally what we measure, even when we are measuring dimensionful quantities. We always measure a physical quantity against a like dimensioned standard. When one measures a length with a ruler or tape-measure, that person is actually counting tick marks on the ruler (a standard used to measure length) and the net result is a dimensionless number. But when that quantity is expressed as the dimensionless number attached to (multiplying) a dimensionful unit, it becomes, conceptually, a dimensionful quantity.

BASE UNITS OF THE INTERNATIONAL SYSTEM (SI)

The General Conference on Weights and Measures has replaced all but one of the definitions of its base (fundamental)

units based on physical objects (such as standard meter sticks or standard kilogram bars) with physical descriptions of the units based on stable properties of the Universe.

For example, the second, the base unit of time, is now defined as that period of time in which the waves of radiation emitted by cesium atoms, under specified conditions, display exactly 9 192 631 770 cycles. The meter, the base unit of distance, is defined by stating that the speed of light, a universal physical constant, is exactly 299 792 458 meters per second. These physical definitions allow scientists to reconstruct meter standards or standard clocks anywhere in the world, or even on other planets, without referring to a physical object kept in a vault somewhere.

In fact, the kilogram is the only base unit still defined by a physical object. The International Bureau of Weights and Measures (BIPM) keeps the world's standard kilogram in Paris, and all other weight standards, such as those of Britain and the United States, are weighed against this standard kilogram.

This one physical standard is still used because scientists can weigh objects very accurately. Weight standards in other countries can be adjusted to the Paris standard kilogram with an accuracy of one part per hundred million. So far, no one has figured out how to define the kilogram in any other way that can be reproduced with better accuracy than this. The 21st General Conference on Weights and Measures, It passed a resolution calling on national standards laboratories to press forward with research to "link the fundamental unit of mass to fundamental or atomic constants with a view to a future redefinition of the kilogram." The 22nd General Conference, in 2003, renewed this request. It is possible that the 24th General Conference, in 2007, will make a change in the definition.

Following are the official definitions of the seven base units, as given by BIPM. The links in the first column are to my (possibly) less obscure definitions.

Meter(m)	distance	"The metre is the length of the path travelled by light in vacuum during a time interval of 1/299 792 458 of a second."
Kilogram	(kg)mass	"The kilogram is equal to the mass of the international prototype of the kilogram."
Second (s)	time"	The second is the duration of 9 192 631 770 periods of the radiation corresponding to the transition between the two hyperfine levels of the ground state of the caesium 133 atom."
Ampere (A)	electric current	"The ampere is that constant current which, if maintained in two straight parallel conductors of infinite length, of negligible circular cross-section, and placed 1 metre apart in vacuum, would produce between these conductors a force equal to 2×10^{-7} newton per metre of length."
Kelvin(K)	temperature	"The kelvin is the fraction 1/273.16 of the thermodynamic temperature of the triple point of water."
Mole (mol)	amount of substance	"The mole is the amount of substance of a system which contains as many elementary entities as there are atoms in 0.012 kilogram of carbon 12. When the mole is used, the elementary entities must be specified and may be atoms, molecules, ions, electrons, other particles, or specified groups of such particles."
Candela (cd)	intensity of light	"The candela is the luminous intensity, in a given direction, of a source that emits monochromatic radiation of frequency 540×10^{12} hertz and that has a radiant intensity in that direction of 1/683 watt per steradian."

DERIVED UNITS OF THE INTERNATIONAL SYSTEM (SI)

In addition to the seven base units, the General Conference on Weights and Measures has approved 22 names for SI units defined as combinations of the base units. Within the SI, these units are the *named derived units.*

Following are links to the definitions of these 22 units. The third column shows how each unit is derived from the preceding units, and the fourth column gives the formal equivalence of the unit in terms of the base units.

Derived Unit	Measures	Derivation	Formal Definition
hertz (Hz)	frequency	/s	s^{-1}
newton (N)	force	$kg \cdot (m/s^2)$	$kg \cdot m \cdot s^{-2}$
pascal (Pa)	pressure	N/m^2	$kg \cdot m^{-1} \cdot s^{-2}$
joule (J)	energy or work	$N \cdot m$	$kg \cdot m^2 \cdot s^{-2}$
watt (W)	power	J/s	$kg \cdot m^2 \cdot s^{-3}$
coulomb (C)	electric charge	$A \cdot s$	$A \cdot s$
volt (V)	electric potential	W/A	$kg \cdot m^2 \cdot s^{-3} \cdot A^{-1}$
farad (F)	electric capacitance	C/V	$kg^{-1} \cdot m^{-2} \cdot s^4 \cdot A^2$
ohm (omega)	electric resistance	V/A	$kg \cdot m^2 \cdot s^{-3} \cdot A^{-2}$
siemens (S)	electric conductance	A/V	$kg^{-1} \cdot m^{-2} \cdot s^3 \cdot A^2$
weber (Wb)	magnetic flux	$V \cdot s$	$kg \cdot m^2 \cdot s^{-2} \cdot A^{-1}$
tesla (T)	magnetic flux density	Wb/m^2	$kg \cdot s^{-2} \cdot A^{-1}$
henry (H)	inductance	Wb/A	$kg \cdot m^2 \cdot s^{-2} \cdot A^{-2}$
degree Celsius (°C)	temperature	K - 273.15	K
radian (rad)	plane angle		$m \cdot m^{-1}$
steradian (sr)	solid angle		$m^2 \cdot m^{-2}$
lumen (lm)	luminous flux	$cd \cdot sr$	$cd \cdot sr$
lux (lx)	illuminance	lm/m^2	$m^{-2} \cdot cd \cdot sr^{-1}$
becquerel(Bq)	activity	/s	s^{-1}
gray (Gy)	absorbed dose	J/kg	$m^2 \cdot s^{-2}$
sievert (Sv)	dose equivalent	Gy (multiplier)	$m^2 \cdot s^{-2}$
katal (kat)	catalytic activity	mol/s	$mol \cdot s^{-1}$

The term *derived unit* covers any algebraic combination of the base units, but it is only the 22 combinations listed above that have approved special names. For example, the SI derived unit of momentum (mass times velocity) has no special name; momentum is stated in kilogram meters per second ($kg \cdot m/s$)

or in newton seconds (N·s). A few SI derived units do have special names that have been defined but not approved. Here are some examples:

Derived Unit	Measures	Derivation
Stere (st)	volume	m^3
Diopter (dpt)	refractive power	m^{-1}
Thermal ohm	thermal resistance	K/W
Poiseuille (Pl)	dynamic viscosity	Pa·s
Rayl	sound impedance	Pa·s/m
Acoustic ohm	sound resistance	Pa·s/m^3
Daraf	electric elastance	F^{-1}
Talbot	luminous energy	lm·s
Nit (nt)	luminance	cd/m^2
Molal	chemical concentration	mol/kg

Metric Prefixes

To help the SI units apply to a wide range of phenomena, the 19th General Conference on Weights and Measures extended the list of metric prefixes so that it reaches from yotta- at 10^{24} (one septillion) to yocto- at 10^{-24} (one septillionth). Here are the metric prefixes, with their numerical equivalents stated in the American system for naming large numbers:

yotta- (Y-)	10^{24}	1 septillion
zetta- (Z-)	10^{21}	1 sextillion
exa- (E-)	10^{18}	1 quintillion
peta- (P-)	10^{15}	1 quadrillion
tera- (T-)	10^{12}	1 trillion
giga- (G-)	10^{9}	1 billion
mega- (M-)	10^{6}	1 million
kilo- (k-)	10^{3}	1 thousand
hecto- (h-)	10^{2}	1 hundred
deka- (da-)**	10	1 ten
deci- (d-)	10^{-1}	1 tenth
centi- (c-)	10^{-2}	1 hundredth
milli- (m-)	10^{-3}	1 thousandth
micro- (μ-)	10^{-6}	1 millionth
nano- (n-)	10^{-9}	1 billionth

pico- (p-)	10^{-12}	1 trillionth
femto- (f-)	10^{-15}	1 quadrillionth
atto- (a-)	10^{-18}	1 quintillionth
zepto- (z-)	10^{-21}	1 sextillionth
yocto- (y-)	10^{-24}	1 septillionth

Notes: It is often asked about prefixes for other multiples, such as 10^4, 10^5, 10^{-4}, and 10^{-5}. The prefix myria- (my-) was formerly used for 10^4, but it is now considered obsolete and it is not accepted in the SI. No prefixes were ever accepted generally for 10^5, 10^{-4}, or 10^{-5}.

There is a widespread misconception that prefixes for positive powers of ten are all capitalized, leading to the use of K- for kilo- and D- for deca-. Although this does seem like a useful idea, it is not correct.

**The *SI Brochure* spelling of this prefix is deca-, but the U.S. National Institute of Standards and Technology (NIST) recommends deka-. National variations in spelling of the prefixes are allowed by the SI. In Italian, for example, hecto- is spelled etto- and kilo- is spelled chilo-. The symbols, however, are the same in all languages, so dam (not dkm) is the symbol for the dekameter and km is the symbol for the Italian chilometro.

The prefixes hecto-, deka-, deci-, and centi- are widely used in everyday life but are generally avoided in scientific work. Contrary to the belief of some scientists, however, the SI does allow use of these prefixes.

The last letter of a prefix is often omitted if the first letter of the unit name is a vowel, causing the combination to be hard to pronounce otherwise. Thus 100 ares is a hectare and 1 million ohms is a megohm. However, the last letter of the prefix is not omitted if pronunciation is not a problem, as in the case of the milliampere. The letter "l" is sometimes added to prefixes before the erg, so 1 million ergs is a megalerg (sounds odd, but better than "megerg").

Binary Prefixes

In computing, a custom arose of using the metric prefixes to specify powers of 2. For example, a kilobit is usually 2^{10} = 1024 bits instead of 1000 bits. This practice leads to considerable confusion. In an effort to eliminate this confusion, in 1998 the International Electrotechnical Commission approved new prefixes for the powers of 2. These prefixes are as follows:

kibi-	Ki-	2^{10} = 1 024
mebi-	Mi-	2^{20} = 1 048 576
gibi-	Gi-	2^{30} = 1 073 741 824
tebi-	Ti-	2^{40} = 1 099 511 627 776
pebi-	Pi-	2^{50} = 1 125 899 906 842 624
exbi-	Ei-	2^{60} = 1 152 921 504 606 846 976

The Commission's ruling is that the metric prefixes should be used in computing just as they are used in other fields. Thus, 5 gigabytes (GB) should mean exactly 5 000 000 000 bytes, and 5 gibibytes (GiB) should mean exactly 5 368 709 120 bytes.

The fate of this innovation is uncertain. So far, very few people are using the IEC binary prefixes. Searches for them on the Internet turn up, for the most part, complaints by people who *don't* want to use them.

CGS AND MKS UNITS

Scientists have adopted the metric system to simplify their calculations and promote communication across national boundaries. However, there have been two ideas as to which metric units should be preferred in science. Scientists working in laboratories, dealing with small quantities and distances, preferred to measure distance in centimeters and mass in grams. Scientists and engineers working in larger contexts preferred larger units: meters for distance and kilograms for mass. Everyone agreed that units of other quantities such as force, pressure, work, power, and so on should be related in a simple way to the basic units, but which basic units should be used?

The result was two clusterings of metric units in science and engineering. One cluster, based on the centimeter, the gram, and the second, is called the CGS system. The other, based on the meter, kilogram, and second, is called the MKS system.

When we say, for example, that the dyne is the CGS unit of force, this determines its definition: it is the force which accelerates a mass of one gram at the rate of one centimeter per second per second. The MKS unit of force, the newton, is the force which accelerates a mass of one kilogram at the rate of one meter per second per second. The ratio between a CGS unit and the corresponding MKS unit is usually a power of 10. A newton accelerates a mass 1000 times greater than a dyne does, and it does so at a rate 100 times greater, so there are 100 000 = 10^5 dynes in a newton.

The CGS system was introduced formally by the British Association for the Advancement of Science in 1874. It found almost immediate favour with working scientists, and it was the system most commonly used in scientific work for many years. Meanwhile, the further development of the metric system was based on meter and kilogram standards created and distributed in 1889 by the International Bureau of Weights and Measures (BIPM). During the 20th century, metric units based on the meter and kilogram — the MKS units — were used more and more in commercial transactions, engineering, and other practical areas. By 1950 there was some discomfort among users of metric units, because the need to translate between CGS and MKS units went against the metric ideal of a universal measuring system. In other words, a choice needed to be made.

In 1954, the Tenth General Conference on Weights and Measures (CGPM) adopted the meter, kilogram, second, ampere, degree Kelvin, and candela as the basic units for all international weights and measures, and in 1960 the Eleventh General Conference adopted the name International System of Units (SI) for this collection of units. (The "degree Kelvin" became the kelvin in 1967.) In effect, these decisions gave the central core of the MKS system preference over the CGS system. Although some of the CGS units remain in use for a variety of purposes, they are being replaced gradually by the SI units selected from the MKS system.

Following is a table of CGS units with their SI equivalents. Note that in some cases there is more than one name for the same unit. The CGS electromagnetic and electrostatic units are not included in this table, except for those which have special names.

CGS unit	Measuring	SI equivalent
barye (ba)	pressure	0.1 pascal (Pa)
biot (Bi)	electric current	10 amperes (A)
calorie (cal)	heat energy	4.1868 joule (J)
darcy	permeability	0.98692×10^{-12} square meter (m^2)
debye (D)	electric dipole moment	3.33564×10^{-30} coulomb meter (C·m)
dyne (dyn)	force	10^{-5} newton (N)
emu	magnetic dipole moment	0.001 ampere square meter ($A \cdot m^2$)
erg	work, energy	10^{-7} joule (J)
franklin (Fr)	electric charge	3.3356×10^{-10} coulomb (C)
galileo (Gal)	acceleration	0.01 meter per second squared ($m \cdot s^{-2}$)
gauss (G)	magnetic flux density	10^{-4} tesla (T)
gilbert (Gi)	magnetomotive force	0.795 775 ampere-turns (A)
kayser (K)	wave number	100 per meter (m^{-1})
lambert (Lb)	luminance	3183.099 candelas per square meter ($cd \cdot m^{-2}$)
langley	heat transmission	41.84 kilojoules per square meter ($kJ \cdot m^{-2}$)
line (li)	magnetic flux	10^{-8} weber (Wb)
maxwell (Mx)	magnetic flux	10^{-8} weber (Wb)
oersted (Oe)	magnetic field strength	79.577 472 ampere-turns per meter ($A \cdot m^{-1}$)
phot (ph)	illumination	10^4 lux (lx)
poise (P)	dynamic viscosity	0.1 pascal second (Pa·s)
stilb (sb)	luminance	10^4 candelas per square meter ($cd \cdot m^{-2}$)
stokes (St)	kinematic viscosity	10^{-4} square meters per second ($m^2 \cdot s^{-1}$)
unit pole	magnetic flux	$1.256\ 637 \times 10^{-7}$ weber (Wb)

PHYSICAL-DIMENSION

All physical quantities are described by parameters such as length, mass and so on, all of which have dimensions. In general we will denote the dimension of length by L, of time by $T$and of mass by M. The dimension of a quantity x will be denoted by square brackets as $[x]$. For example, the dimension of the volume of an object V is denoted by $[V] = L \times L \times L = L^3$. An equation such at $X + Y = Z$ is consistent only if the dimension of both the sides match, in this case only if all the three quantities X, Y and Z have the same dimensions. Checking that the dimensions agree for an equations is one of the first steps that one should take when analyzing a new set of equations. Dimensional analysis can lead to fairly sophisticated results; however, dimensional analysis can only lead to the general form of the equation, and cannot fully specify the solution.

For example, if one has two dimensional quantities X and Y, and we know that the solution A is a dimensionless combination of X and Y, then we can straight away conclude that the solution must be a function only of the ratio X/Y; however, based only on dimensional considerations the solution can be either A = 1 + X/Y or equivalently A = 100 - Y/X, and hence the complete answer for A cannot simply be obtained from dimensional analysis. The units that one uses for the various dimensional quantities are specific choices of what is one unit of length, time and so on. Units are arbitrary, and any set of consistent units can be used. For example, we can use meters or equivalently inches to measure length, grams or pounds to measure mass, and so on. Of course one can always convert from one set of units to another, and is analogous to converting from one currency to another. We will use the SI (Systeme International)units, which previously was known as the MKS system of units. In these units distance is measured in meters (m), mass is measured in terms of kilogram (kg), and time is measured in seconds (s). Hence the dimension of velocity is given by $[v] = LT^{-1}$, and in SI units we have $[v] = ms^{-1}$.

DIMENSIONS OF PHYSICAL QUANTITIES

We are familiar with dimensions of motion and motion related quantities. Often, we specify the description of physical process by numbers of coordinates involved - one, two or three. It indicates the context of motion in space. The dimension of physical quantities follows the same philosophy and indicates the nature of the constitution of quantities. In other words, dimension of a physical quantity indicates how it relates to one of the seven basic/ fundamental quantities. Basic quantities are the seven dimensions of the physical quantities.

Dimensions of a physical quantity are the powers with which basic quantities are raised to represent it. The dimension of a physical quantity in an individual basic quantity is the power with which basic quantity is raised in the representation of physical quantity.

We should be clear here that the dimension is not merely a power, but a combination of basic quantity and its power.

A pair of square bracket is used to represent the dimension of individual basic quantity with its symbol enclosed within the bracket. There is a convention in using symbol of basic quantities. The dimensions of seven basic quantities are represented as:

1. Mass: [M]
2. Length: [L]
3. Time: [T]
4. Current: [A]
5. Temperature: [K]
6. Amount of substance: [mol]
7. Luminous Intensity: [cd]

We can see here that there is no pattern. Sometimes we use initial letter of the basic physical quantity, and sometimes we use initial letter of basic unit and even abbreviated name of the basic unit like "mol".

DIMENSIONS OF DERIVED QUANTITIES

The dimensions of derived quantities include all dimensions in individual basic quantities. In order to understand the technique to write dimensions of a derived quantity, we consider the case of force. The force is defined as:

$F = ma$

Thus, dimensions of force is:

$$\Rightarrow [F] = [M][\alpha]$$

The dimension of acceleration, represented as [a], is itself a derived quantity being the ratio of velocity and time. In turn, velocity is also a derived quantity, being ratio of length and time.

$$\Rightarrow [F] = [M][\alpha] = [M]\left[vT^{-1}\right] = [M]\left[LT^{-1}T^{-1}\right] = \left[MLT^{-2}\right]$$

We read the dimension of force as: it has "1" dimension in mass, "1" dimension in length and "-2" dimension in time. This reading emphasizes the fact that the dimension is not merely a power, but a combination of basic quantity and its power.

DIMENSIONAL FORMULA

The expression of dimensional representation is also called "dimensional formula" of the given physical quantity.

For brevity, we do not include all the seven basic dimensions in the dimensional formula.

Force, $[F] = [MLT^{-2}]$

Velocity, $[V] = [LT^{-1}]$

Charge, $[q] = [AT]$

Specific heat, $[s] = [L^2T^2K^{-1}]$

Gas constant, $[R] = [ML^2T^{-2}K^{-1}mol^{-1}]$

It is conventional to omit power of "1". Further, in mechanics, we can optionally use all three symbols corresponding to three basic quantities, which may be involved. Even if one or two of them are not present, it is considered conventional to report absence of the dimension in a particular basic quantity. It is done by raising the symbol to zero as:

$$[v] = [M^0LT^{-1}]$$

It should be noted here that dimensional representation of a physical quantity does not include "magnitude (n)" of the physical quantity. Further, dimensional formula does not distinguish nature of quantity. For example, the nature of force (or types of force) has no bearing on its dimensional representation.

It, then, also follows that variants of a physical quantity bears the same dimensional representation. For example, velocity - whether instantaneous, average, relative - has the same dimensional representation $[LT^{-1}]$ Moreover, dimensional representation of a vector and its scalar counterpart is same. For example, the dimensions of velocity and speed are same. Further, dimensions of the difference of a physical quantity are same as that of physical quantity itself. For this reason, dimensions of velocity and difference of two velocities are same.

A physical quantity need not have dimensions in any of the basic quantities. Such is the case, where physical quantities are equal to the ratio of quantities having same dimensions. Take the case of an angle, which is a ratio of arc(length) and radius(length). The dimensions of such physical quantity are zero in each of the basic quantities.

$$\text{Angles, } [\theta] = [M^0L^0T^0]$$

Similar is the case with Reynold's number, used in fluid mechanics. It also does not have dimensions in basic quantities. Thus, a physical quantity can be either "dimensional" or "dimensionless".

On the other hand, there are numerical constants like trigonometric function, pi etc, which are not dependent on the basic physical quantities. The dimensions of such a constant are zero in each of the basic quantities. They are, therefore, called "dimensionless" constants.

$$\text{pi}, [\pi] = [M^0L^0T^0]$$

However, there are physical constants, which appear as the constant of proportionality in physical formula. These constants have dimensions in basic quantities. Such constant like Gravitational constant, Boltzmann constant, Planck's constant etc. are, therefore, "dimensional" constant.

In the nutshell, we categorize variables and constants in following categories:

- Dimensional variable: speed, force, current etc.
- Dimensionless variable: angle, reynold's number etc.
- Dimensional constant: universal gas constant, permittivity, permeability
- Dimensionless constant: numerical constants, mathematical constants, trigonometric functions etc.

DIMENSIONAL EQUATION

Dimensional equation is obtained by equating dimension of a given physical quantity with its dimensional formula. Hence, followings constitute a dimensional equation:

$$[F] = [MLT^{-2}]$$
$$[v] = [M^0LT^{-1}]$$

DIMENSIONAL ANALYSIS

Dimensional analysis gives an insight into the composition of a physical quantity. Take the case of frequency. Its SI unit is Hertz. What is its true relation with basic quantities? Notwithstanding the exact definition, its dimensional formula tells us that it is actually inverse of time (T). It has the dimension of "-1" in time.

In this section, we shall discuss three important applications of dimensional analysis:

1. Conversion of units from one system to another
2. Checking a dimensional formula and
3. Driving a formula.

CONVERSION FROM ONE SYSTEM TO ANOTHER

Working of conversion of units combines the concept of unit with the concept of dimensional formula. We know that:

$$Q = n_1 u_1 = n_2 u_2$$

Since unit represents a standard division of the quantity, the dimensional formula of the unit quantity is same as that of the quantity in question. Let dimensions of units in two systems are:

$$u_1 = \left[M_1^a L_1^b T_1^c\right]$$

$$u_2 = \left[M_2^a L_2^b T_2^c\right]$$

Combining equations, we have:

$$\Rightarrow n_1\left[M_1^a L_1^b T_1^c\right] = n_2\left[M_2^a L_2^b T_2^c\right]$$

$$\Rightarrow n_2 = \frac{\left[M_1^a L_1^b T_1^c\right]}{\left[M_2^a L_2^b T_2^c\right]} n_1$$

This is the formula used to convert a quantity from one system of units to another. In case, only 1 unit of the quantity in first system is involved, then we can put [$n_1 = 1$] in the above equation..

Example

Problem 1: Convert 1 Joule in SI system into ergs in "cgs" system

Solution: Putting $n_1 = 1$ and applying formula, we have:

$$\Rightarrow n_2 = \frac{\left[M_1^a L_1^b T_1^c\right]}{\left[M_2^a L_2^b T_2^c\right]}$$

Now, the dimensional formula of energy is [$M L^2 T^{-2}$]. Hence, a = 1, b = 2 and c = -2. Putting values in the equation and rearranging,

$$\Rightarrow n_2 = \left[\frac{1kg}{1gm}\right]^1 \times \left[\frac{1m}{1cm}\right]^2 \times \left[\frac{1s}{1s}\right]^{-2}$$

The basic units of SI system is related to corresponding basic units in "cgs" system as

$1kg = 10^3 gm$

$1m = 10^2 cm$

$1s = 1s$

Putting in the conversion formula, we have:

$$\Rightarrow n_2 = \left[\frac{10^3 gm}{1gm}\right]^1 \times \left[\frac{10^2 cm}{1cm}\right]^2 \times \left[\frac{1s}{1s}\right]^{-2}$$

$$\Rightarrow n_2 = 10^3 \times 10^4 = 10^7$$

Hence,

1J = 107 ergs

CHECKING A FORMULA

The terms of an equation connecting different physical quantities are dimensionally compatible. Each term of the equation are equal in dimensions. This statement is known as "principle of homogeneity of dimensions".

Hence, a formula not having dimensionally compatible terms are incorrect. Therefore, dimensions on two sides of a dimensional equation are equal. Further, it also means that terms connected with plus or minus sign in a dimensional formula should have the same dimensions. This fact, as a matter of fact, can be used to determine the nature of constants/ variables and their units as illustrated in the example here.

Example

Problem 2: The Vander Wall's equation is given as:

$$\left(P + \frac{a}{V^2}\right)(V - b) = RT$$

where "P" is pressure, "V" is volume, "T" is temperature and "R" is gas constant. What are the units of constants "a" and "b"?

Solution: Applying principle of homogeneity of dimensions, we realise that dimensions of the terms connected with "plus" or "minus" signs are equal. Hence,

$$\Rightarrow \left[\frac{a}{V^2}\right] = [P] = \left[\frac{F}{A}\right] = \left[\frac{MLT^{-2}}{L^2}\right] = \left[ML^{-1}T^{-2}\right]$$

$$\Rightarrow [a] = \left[ML^{-1}T^{-2}\right]\left[V^2\right] = \left[ML^{-1}T^{-2}\right]\left[L^3\right]^2$$

$$\Rightarrow [a] = \left[ML^5T^{-2}\right]$$

The unit of "a" is " kg - m^5/s^2".

Also, $\Rightarrow [b] = [V] = \left[L^3\right]$

The unit of "b" is same as that of volume i.e. " m^3 "

DERIVING A FORMULA

This is a curious aspect of the application of dimensional analysis. If we could have the general capability to construct formula in this manner, then we would have known all the secrets of nature without much difficulty. Clearly, derivation of formula by dimensional analysis is possible under certain limited circumstance only. Even then, significance of deriving formula is important as it has contributed quite remarkably in fluid mechanics and other branches of science.

We shall use dimensional analysis to derive Stoke' law for viscous force in the example given here. We should appreciate, however, that dimensional analysis can not determine constant of a formula. As such, we shall work the example given here with this particular limitation.

Example

Problem 3: Stoke observed that viscous drag force, "F", depends on

(i) Coefficient of viscosity, "η"
(ii) Velocity of the spherical mass, "v" and
(iii) Radius of the spherical body, "r".

If the dimensional formula of coefficient of viscosity $\left[ML^{-1}T^{-1}\right]$ is drive Stoke's law for viscous force on a small spherical body in motion through a static fluid medium.

Solution: We first write the observations in mathematical form as:

$F \alpha \eta^{a}; F \alpha v^{b}; F \alpha r^{c}$

Combining, we have:

$$F = K\eta^{a} v^{b} r^{c}$$

where "K" is the constant of proportionality and is dimensionless. Applying principle of homogeneity of dimensions, the dimensions on the left and right side of the equation should be same. Hence,

$$[F] = \left[K\eta^{a} v^{b} r^{c}\right]$$

Neglecting "K" as it is dimensionless and putting dimensional formulas of each quantity, we have:

$$[MLT^{-2}] = [ML^{-1}T^{-1}]^{a} \times [LT^{-1}]^{b} \times [L]^{c}$$

Expanding powers and rearranging, we have:

$$[MLT^{-2}] = [M^{a}L^{-a+b+c}\,T^{-a-b}]$$

For dimensional consistency as required by the principle, the powers of individual basic quantities should be equal. Hence,

$$\Rightarrow a = 1$$

$$\Rightarrow -a + b + c = 1$$

$$\Rightarrow -a - b = -2$$

Combining first and second equations, we have:

$$\Rightarrow -1 + b + c = 1$$

$$\Rightarrow b + c =$$

Combining first and third equations, we have:

$$\Rightarrow -1 - b = -2$$

$$\Rightarrow b = 1$$

Hence, c = 1. Now putting values in the equation for the force,

$$F = k\eta vr$$

Note: It is evident that we can not determine the constant, "K", using dimensional analysis. It is found experimentally to be equal to "6π". Hence, Stoke's law for viscous drag is written as:

$$\Rightarrow F = 6\pi\eta vr$$

This example highlights the limitation of dimensional analysis in certain respect in driving formula. First, we notice that it is not possible to evaluate constant by dimensional analysis. It is also clear that dependent observations or relations should be less than or equal to the basic quantities involved in the quantity whose expression is being derived.

The numbers of basic quantities determine the numbers of equations available for simultaneous evaluation. As such, it is required that the numbers of quantities should be limited within the numbers of basic quantities. If the quantity pertains to mechanics, then numbers of quantities should be limited to "3".

It can also be seen that dimensional analysis can not be applied in situation, where there are more than one term like in the case of $x = ut + \frac{1}{2}at^2$. We can not derive this equation as dimensional analysis will yield one of two terms - not both simultaneously.

DIMENSIONLESS QUANTITY

In dimensional analysis, a dimensionless quantity (or more precisely, a quantity with the dimensions of 1) is a quantity without any physical units and thus a pure number. Such a number is typically defined as a product or ratio of quantities which do have units, in such a way that all units cancel.

Examples

"Out of every 10 apples I gather, 1 is rotten." — the rotten-to-gathered ratio is (1 apple) / (10 apples) = 0.1 = 10%, which is a dimensionless quantity. Another more typical example in physics and engineering is the measure of plane angles with the unit of "radian". An angle measured this way is the length of arc lying on a circle (with centre being the vertex of the angle) swept out by the angle to the length of the radius of the circle. The units of the ratio is length divided by length which is dimensionless.

Dimensionless quantities are widely used in the fields of mathematics, physics, engineering, and economics but also in everyday life. Whenever one measures any physical quantity, they are measuring that physical quantity against a like dimensioned standard. Whenever one commonly measures a length with a ruler or tape measure, they are counting tick marks on the standard of length they are using, which is a dimensionless number. When they attach that dimensionless number (the number of tick marks) to the units that the standard represents, they *conceptually* are referring to a dimensionful quantity. A quantity Q is defined as the product of that dimensionless number n (the number of tick marks) and the unit U (the standard):

$$Q \underset{=}{def} nU$$

But, ultimately, people always work with dimensionless numbers in reading measuring instruments and manipulating (changing or calculating with) even dimensionful quantities.

In case of dimensionless quantities the unit U is a quotient of like dimensioned quantities that can be reduced to a number (kg/kg = 1, µg/g = 1e-6). Dimensionless quantities can also carry dimensionless units like % (=0.01), ppt (=1e – 3), ppm (=1e – 6), ppb (=1e – 9).

The CIPM Consultative Committee for Units toyed with the idea of defining the unit of 1 as the 'uno', but the idea was dropped.

Properties

- A dimensionless quantity has no physical unit associated with it. However, it is sometimes helpful to use the same units in both the numerator and denominator, such as kg/kg, to show the quantity being measured.
- A dimensionless proportion has the same value regardless of the measurement units used to calculate it. It has the same value whether it was calculated using the SI system of units or the imperial system of units. This doesn't hold for all dimensionless quantities; it is guaranteed to hold only for proportions.

BUCKINGHAM π-THEOREM

According to the Buckingham π-theorem of dimensional analysis, the functional dependence between a certain number (e.g., n) of variables can be reduced by the number (e.g., k) of independent dimensions occurring in those variables to give a set of $p = n$ " k independent, dimensionless quantity. For the purposes of the experimenter, different systems which share the same description by dimensionless quantity are equivalent.

Example

The power consumption of a stirrer with a particular geometry is a function of the density and the viscosity of the fluid to be stirred, the size of the stirrer given by its diameter, and the speed of the stirrer. Therefore, we have $n = 5$ variables representing our example.

Those $n = 5$ variables are built up from $k = 3$ dimensions which are:

- Length: L (m)
- Time: T (s)
- Mass: M (kg)

According to the π-theorem, the $n = 5$ variables can be reduced by the $k = 3$ dimensions to form $p = n - k = 5 - 3 = 2$ independent dimensionless numbers which are in case of the stirrer

- Reynolds number (This is the most important dimensionless number; it describes the fluid flow regime)
- Power number (describes the stirrer and also involves the density of the fluid)

EQUATIONS AND DIMENSIONLESS VARIABLES

The equations of motion of particles in expanding coordinates, which are used in our PM code,. Different numerical effects (including resolution). We use comoving coordinates ofr particles $x = x\ (t)$, which are related to proper coordinates by $r = a(t)$x, where $a(t) = (1 + z)^{-1}$ is the expansion parameter. Instead of using peculiar velocity $v_{pec} = ax$ we write the equations for particle momenta p:

$$p = a^2x,\ v_{pec:} = p/a$$

This choice of "velocity" simplifies the equations of motion by removing a few terms with $\dot{a}/a$. It is also convenient to change the time variable from time t to the expansion parameter a. The equations governing the motion of particles are:

$$\frac{dp}{da} = -\frac{\nabla\phi}{\dot{a}}, \frac{dx^{\cdot}}{da} = \frac{p}{\dot{a}a^2}$$

$$\nabla^2\phi = 4\pi G\Omega_{m(t)}a^{2px}(t)\delta$$

$$= 4\pi G\Omega_0 p_{a,0}\frac{\delta}{a}, \delta = \frac{p(x) - pb}{pb}$$

$$\dot{a}^{\sqrt{a}} = H_0\sqrt{\Omega_0 + \Omega_{curv,0}a + \Omega_{A,0a^3}} = \Omega_0 + \Omega_{curv,0} + \Omega_{A,0} = 1$$

where $\Omega_0 = \Omega_{m(z=0)}, \Omega_{curv,0}$ and $\Omega_{A,0}$ are the density of the matter, effective density of the curvature, and the cosmological constant in units of the critical density at $z = 0$. The curvature contribution is positive for negative curvature.

Dimensionless variables (shown with tildas below) are defined by introducing the length of a cell of the grid x_0 and by measuring the time in units of $1/H_0$:

$$x = x_0\tilde{x}, t = \tilde{t}/H_0,$$

$$v_{pec} = (x_0H_0)\tilde{p}/a, \phi = \tilde{\phi}(x_0H_0)^2$$

$$\text{p} = \frac{\tilde{p}}{a^3}\frac{3H_0^2}{8\pi G}\Omega_0$$

Equations can be rewritten in terms of dimensionless variables:

$$\frac{d\tilde{p}}{da} = -F(a)\tilde{\nabla}\tilde{\phi}, \frac{d\tilde{x}}{da} = F(a)\frac{\tilde{P}}{a^2}$$

$$\tilde{\nabla}^2\tilde{\phi} = \frac{3}{2}\frac{\Omega_0}{a}\tilde{p}$$

where

$$F(a) \equiv H_0/\dot{a} = \frac{\left(\Omega_0 + \Omega_{curv,0a} + \Omega_{\Lambda,0a^3}\right)}{a}-1/2$$

Equations are solved numerically by the PM code.

If L is the length of the computational box at $z = 0$, N_{grid} is the number of grid cells in one direction, and N_{row} is the number of particles in one direction, which contribute a fraction Ω_0 of the critical density, then the transformations from dimensionless variables given by the code to dimensional variables are given by

$$x = x_0\tilde{x}, x_0 = \frac{L}{N_{grid}} 7.8kpc\left(\frac{L_{Mpc}}{N_{grid}/128}\right),$$

$$v_{pec} = (x_0H_0)\frac{\tilde{p}}{a} = 0.781\frac{km}{s}\cdot\frac{\tilde{p}}{a}\cdot\frac{L_{Mpch}}{N_{grid}/128},$$

$Mass = N_{\text{particles.}}\ m_1,\ m_1 =$

$$\Omega_0 p_{a,0}\left(\frac{L}{N_{row}}\right)3 = 1.32\cdot 10^5\left(\Omega_0 h^2\right)\left(\frac{L_{Mpc}}{N_{row}/128}\right)_3$$

LIST OF DIMENSIONLESS QUANTITIES

There are infinitely many dimensionless quantities and they are often called numbers. Some of those that are used most often have been given names, as in the following list of examples (alphabetical order):

Name	Field of application
Abbe number	optics (dispersion in optical materials)
Albedo	climatology, astronomy (reflectivity of surfaces or bodies)
Archimedes number	motion of fluids due to density differences
Bagnold number	flow of grain, sand, etc.
Biot number	surface vs. volume conductivity of solids
Bodenstein number	residence-time distribution
Bond number	capillary action driven by buoyancy
Brinkman number	heat transfer by conduction from the wall to a viscous fluid
Brownell Katz number	combination of capillary number and Bond number
Capillary number	fluid flow influenced by surface tension
Coefficient of static friction	friction of solid bodies at rest
Coefficient of kinetic friction	friction of solid bodies in translational motion
Colburn j factor	dimensionless heat transfer coefficient
Courant-Friedrich-Levy number	non-hydrostatic dynamics
Damköhler numbers	reaction time scales vs. transport phenomena
Darcy friction factor	fluid flow
Dean number	vortices in curved ducts
Deborah number	rheology of viscoelastic fluids
Drag coefficient	flow resistance
Eckert number	convective heat transfer
Ekman number	geophysics (frictional (viscous) forces)
Elasticity (economics)	widely used to measure how demand or supply responds to price changes

Eötvös number	determination of bubble/drop shape
Euler number	hydrodynamics (pressure forces vs. inertia forces)
Fanning friction factor	fluid flow in pipes
Feigenbaum constants	chaos theory (period doubling)
Fine structure constant	quantum electrodynamics (QED)
Foppl–von Karman number	thin-shell buckling
Fourier number	heat transfer
Fresnel number	slit diffraction
Froude number	wave and surface behaviour
Gain	electronics (signal output to signal input)
Galilei number	gravity-driven viscous flow
Graetz number	heat flow
Grashof number	free convection
Hagen number	forced convection
Karlovitz number	turbulent combustion
Knudsen number	continuum approximation in fluids
Kt/V	medicine
Laplace number	free convection within immiscible fluids
Lewis number	ratio of mass diffusivity and thermal diffusivity
Lockhart-Martinelli parameter	flow of wet gases
Lift coefficient	lift available from an airfoil at a given angle of attack
Mach number	gas dynamics
Magnetic Reynolds number	magnetohydrodynamics
Manning roughness coefficient	open channel flow (flow driven by gravity)
Marangoni number	Marangoni flow due to thermal surface tension deviations
Morton number	determination of bubble/drop shape
Nusselt number	heat transfer with forced convection
Ohnesorge number	atomization of liquids, Marangoni flow
Péclet number	advection–diffusion problems
Peel number	adhesion of microstructures with substrate

Pi	mathematics (ratio of a circle's circumference to its diameter)
Poisson's ratio	elasticity (load in transverse and longitudinal direction)
Power factor	electronics (real power to apparent power)
Power number	power consumption by agitators
Prandtl number	forced and free convection
Pressure coefficient	pressure experienced at a point on an airfoil
Radian	measurement of angles
Rayleigh number	buoyancy and viscous forces in free convection
Refractive index	electromagnetism, optics
Reynolds number	flow behaviour (inertia vs. viscosity)
Richardson number	effect of buoyancy on flow stability
Rockwell scale	mechanical hardness
Rossby number	inertial forces in geophysics
Schmidt number	fluid dynamics (mass transfer and diffusion)
Sherwood number	mass transfer with forced convection
Sommerfeld number	boundary lubrication
Stanton number	heat transfer in forced convection
Stefan number	heat transfer during phase change
Stokes number	particle dynamics
Strain	materials science, elasticity
Strouhal number	continuous and pulsating flow
Taylor number	rotating fluid flows
van 't Hoff factor	quantitative analysis (K_f and K_b)
Weaver flame speed number	laminar burning velocity relative to hydrogen gas
Weber number	multiphase flow with strongly curved surfaces
Weissenberg number	viscoelastic flows
Womersley number	continuous and pulsating flows

DIMENSIONLESS PHYSICAL CONSTANTS

Certain physical constants, such as the speed of light in a vacuum, are normalized to 1 if the units for time, length, mass, charge, and temperature are chosen appropriately. The resulting system of units is known as Planck units. However, a handful of dimensionless physical constants cannot be eliminated in any system of units; their values must be determined experimentally. The resulting fundamental physical constants include:

- α, the fine structure constant and the electromagnetic coupling constant
- β, the ratio of the rest mass of the proton to that of the electron
- More generally, the masses of all fundamental particles relative to that of the electron
- The strong Coupling constant
- The gravitational coupling constant

2

Mechanics and Dynamics

SPEED AND VELOCITY

The motion of a moving object can be explained using either Newton's Laws and vector principles or by means of the Work-Energy Theorem. The same concepts and principles used to describe and explain the motion of an object can be used to describe and explain the parabolic motion of a projectile. In this unit, we will see that these same concepts and principles can also be used to describe and explain the motion of objects which either move in circles or can be approximated to be moving in circles. Kinematic concepts and motion principles will be applied to the motion of objects in circles and then extended to analyse the motion of such objects as roller coaster cars, a football player making a *circular* turn, and a planet orbiting the sun. We will see that the beauty and power of physics lies in the fact that a few simple concepts and principles can be used to explain the mechanics of the entire universe.

Suppose that you were driving a car with the steering wheel turned in such a manner that your car followed the path of a perfect circle with a constant radius. And suppose that as you drove, your speedometer maintained a constant reading of 10 mi/hr. In such a situation as this, the motion of your car could be described as experiencing uniform circular motion. Uniform circular motion is the motion of an object in a circle with a constant or uniform speed.

Calculation of the Average Speed

Uniform circular motion - circular motion at a constant speed - is one of many forms of circular motion. An object moving in uniform circular motion would cover the same linear distance in each second of time. When moving in a circle, an object traverses a distance around the perimeter of the circle.. So if your car were to move in a circle with a constant speed of 5 m/s, then the car would travel 5 meters along the perimeter of the circle in each second of time. The distance of one complete cycle around the perimeter of a circle is known as the circumference. With a uniform speed of 5 m/s, a car could make a complete cycle around a circle which had a circumference of 5 meters. At this uniform speed of 5 m/s, each cycle around the 5-m circumference circle would require 1 second. At 5 m/s, a circle with a circumference of 20 meters could be made in 4 seconds; and at this uniform speed, every cycle around the 20-m circumference of the circle would take the same time period of 4 seconds. This relationship between the circumference of a circle, the time to complete one cycle around the circle, and the speed of the object is merely an extension of the average speed equation.

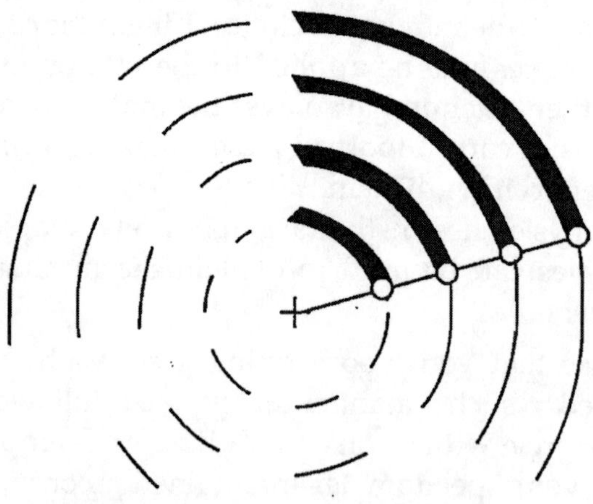

Each bulb on the wine moves around the perimeter of the circle in the same amount of time; but the larger the radius of the circle, the greater the speed.

$$\text{Average Speed} = \frac{\text{distance}}{\text{time}} = \frac{\text{circumference}}{\text{time}}$$

The circumference of any circle can be computed using from the radius according to the equation

Circumference = 2*pi*Radius

Combining these two equations above will lead to a new equation relating the speed of an object moving in uniform circular motion to the radius of the circle and the time to make one cycle around the circle (period).

$$\text{Average Speed} = \frac{2 * \pi * R}{T}$$

where R represents the radius of the circle and T represents the period. This equation, like all equations, can be used as an algebraic recipe for problem solving. It also can be used to guide our thinking about the variables in the equation relate to each other. For instance, the equation suggests that for objects moving around circles of different radius in the same period, the object traversing the circle of larger radius must be traveling with the greatest speed. In fact, the average speed and the radius of the circle are directly proportional. A twofold increase in radius corresponds to a twofold increase in speed; a threefold increase in radius corresponds to a three—fold increase in speed; and so on. To illustrate, consider a strand of four LED lights positioned at various locations along the strand. The strand is held at one end and spun rapidly in a circle. Each LED light traverses a circle of different radius. Yet since they are connected to the same wire, their period of rotation is the same. Subsequently, the LEDs which are further from the centre of the circle are traveling faster in order to sweep out the circumference of the larger circle in the same amount of time. If the room lights are turned off, the LEDs created an arc which could be perceived to be longer for those LEDs which were traveling faster - the LEDs with the greatest radius. This is illustrated in the diagram at the right.

THE DIRECTION OF THE VELOCITY VECTOR

Objects moving in uniform circular motion will have a constant speed. But does this mean that they will have a constant velocity? The Physics Classroom that speed and velocity refer to two distinctly different quantities. Speed is a scalar quantity and velocity is a vector quantity. Velocity, being a vector, has both a magnitude and a direction. The magnitude of the velocity vector is the instantaneous speed of the object. The direction of the velocity vector is directed in the same direction which the object moves. Since an object is moving in a circle, its direction is continuously changing. At one moment, the object is moving northward such that the velocity vector is directed northward. One quarter of a cycle later, the object would be moving eastward such that the velocity vector is directed eastward. As the object *rounds* the circle, the direction of the velocity vector is different than it was the instant before. So while the magnitude of the velocity vector may be constant, the direction of the velocity vector is changing. The best word that can be used to describe the direction of the velocity vector is the word tangential. The direction of the velocity vector at any instant is in the direction of a tangent line drawn to the circle at the object's location. (A tangent line is a line which touches a circle at one point but does not intersect it.) The diagram at the right shows the direction of the velocity vector at four different point for an object moving in a clockwise direction around a circle. While the actual direction of the object (and thus, of the velocity vector) is changing, it's direction is always tangent to the circle.

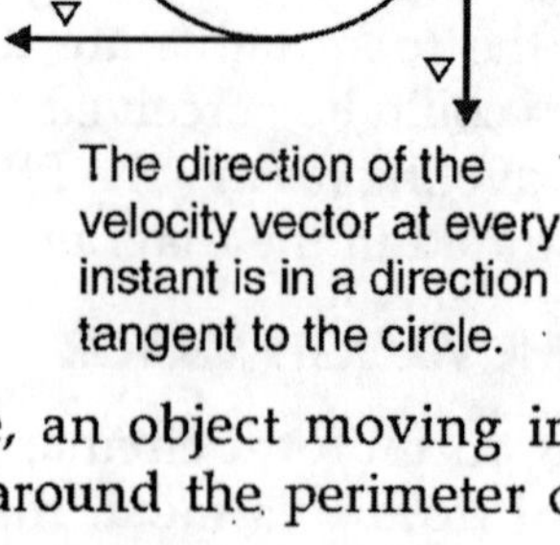

The direction of the velocity vector at every instant is in a direction tangent to the circle.

To summarize, an object moving in uniform circular motion is moving around the perimeter of the circle with a constant speed. While the speed of the object is constant, its velocity is changing. Velocity, being a vector, has a constant magnitude but a changing direction. The direction is always directed tangent to the circle and as the object turns the circle, the tangent line is always pointing in a new direction.

ACCELERATION

An object moving in uniform circular motion is moving in a circle with a uniform or constant speed. The velocity vector is constant in magnitude but changing in direction. Because the speed is constant for such a motion, many students have the misconception that there is no acceleration. "After all," they might say, "if I were driving a car in a circle at a constant speed of 20 mi/hr, then the speed is neither decreasing nor increasing; therefore there must not be an acceleration." At the centre of this common student misconception is the wrong belief that acceleration has to do with speed and not with velocity. But the fact is that an accelerating object is an object which is changing its velocity. And since velocity is a vector which has both magnitude and direction, a change in either the magnitude or the direction constitutes a change in the velocity. For this reason, it can be safely concluded that an object moving in a circle at constant speed is indeed accelerating. It is accelerating because the direction of the velocity vector is changing.

To understand this at a deeper level, we will have to combine the definition of acceleration with a review of some basic vector principles. The Physics Classroom that acceleration as a quantity was defined as the rate at which the velocity of an object changes. As such, it is calculated using the following equation:

$$\text{Ave.acceleration} = \frac{\Delta\,\text{velocity}}{\text{time}} = \frac{v_f - v_i}{t}$$

where v_i represents the initial velocity and v_f represents the final velocity after some time of t. The numerator of the equation is found by subtracting one vector (v_i) from a second vector (v_f). But the addition and subtraction of vectors from each other is done in a manner much different than the addition and subtraction of scalar quantities. Cons er the case of an object moving in a circle about point C as shown in the diagram below. In a time of t seconds, the object has moved from point A to point B. In this time, the velocity has changed from v_i to v_f. The process of subtracting v_i from v_f is shown in the vector diagram; this process yields the change in velocity.

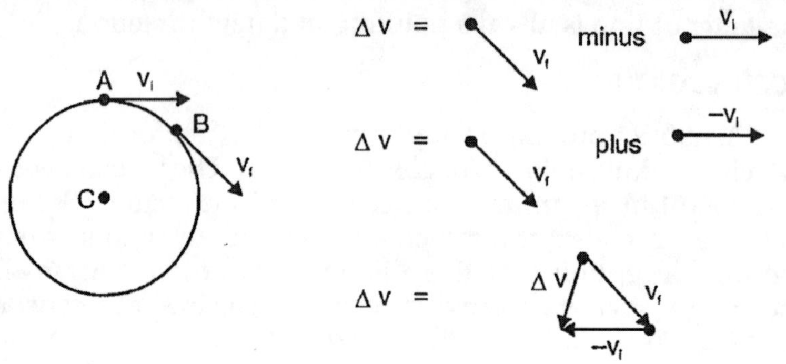

DIRECTION OF THE ACCELERATION VECTOR

Note in the diagram above that there is a velocity change for an object moving in a circle with a constant speed. A careful inspection of the velocity change vector in the above diagram shows that it points down and to the left. At the midpoint along the arc connecting points A and B, the velocity change is directed towards point C - the centre of the circle. The acceleration of the object is dependent upon this velocity change and is in the same direction as this velocity change. The acceleration of the object is in the same direction as the velocity change vector; the acceleration is directed towards point C as well - the centre of the circle. Objects moving in circles at a constant speed accelerate towards the centre of the circle.

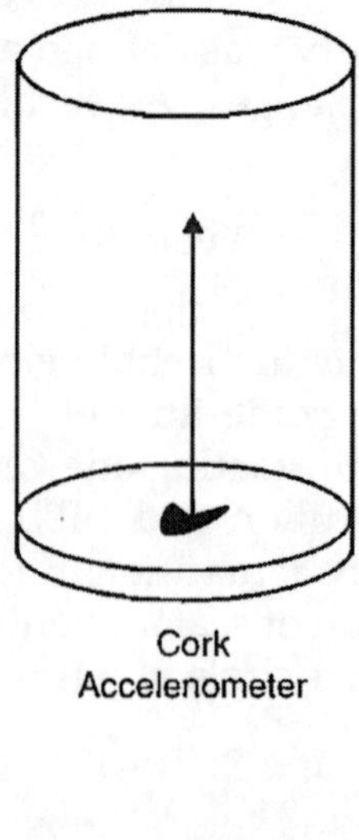

Cork
Accelenometer

The acceleration of an object is often measured using a device known as an accelerometer. A simple accelerometer consists of an object immersed in a fluid such as water. Consider a sealed jar which is filled with water. A cork attached to the lid by a string can serve as an accelerometer. To test the direction of acceleration for an object moving in a circle, the jar can be inverted and attached to the end of a short section of a wooden 2 × 4. A second accelerometer constructed in the same manner can be attached to the opposite end of the 2 × 4. If the 2 × 4 and accelerometers are clamped to a rotating platform and spun in a circle, the direction of the acceleration can be clearly seen by the direction of lean of the corks. As the cork-water combination spins in a circle, the cork leans towards the centre of the circle. The least massive of the two objects always leans in the direction of the acceleration. In the case of the cork and the water, the cork is least massive (on a per mL basis) and thus it experiences the greater acceleration. Having less inertia (owing to its smaller mass on a per mL basis), the cork resists the acceleration the least and thus *leans* to the inside of the jar towards the centre of the circle. This is observable evidence that an object moving in circular motion at constant speed experiences an acceleration which is directed towards the centre of the circle.

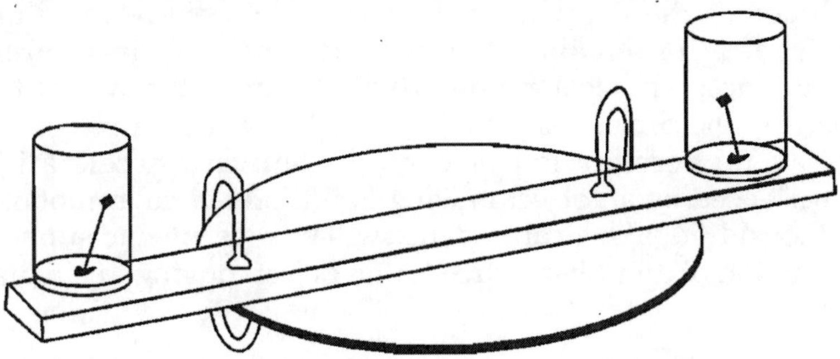

The cork of a water accelerometer points inward towards the center of the circle when placed upon a upon a rotating platform, thus indicating an inward acceleration for circular motion.

Another simple homemade accelerometer involves a lit candle centreed vertically in the middle of an open-air glass.

If the glass is held level and at rest (such that there is no acceleration), then the candle flame extends in an upward direction. However, if you hold the glass-candle system with an outstretched arm and spin in a circle at a constant rate (such that the flame experiences an acceleration), then the candle flame will no longer extend vertically upwards. Instead the flame deflects from its upright position. This signifies that there is an acceleration when the flame moves in a circular path at constant speed. The deflection of the flame will be in the direction of the acceleration. This can be explained by asserting that the hot gases of the flame are less massive (on a per mL basis) and thus have less inertia than the cooler gases which surround. Subsequently, the hotter and lighter gases of the flame experience the greater acceleration and will lean in the direction of the acceleration. A careful examination of the flame reveals that the flame will point towards the centre of the circle, thus indicating that not only is there an acceleration; but that there is an inward acceleration. This is one more piece of observable evidence which indicates that objects moving in a circle at a constant speed experience an acceleration which is directed towards the centre of the circle.

So thus far, we have seen a geometric proof and two real-world demonstrations of this inward acceleration. At this point it becomes the decision of the student to believe or to not believe. Is it sensible that an object moving in a circle experiences an acceleration which is directed towards the centre of the circle? Can you think of a logical reason to believe in say no acceleration or even an outward acceleration experienced by an object moving in uniform circular motion? In the additional logical evidence will be presented to support the notion of an inward force for an object moving in circular motion.

THE CENTRIPETAL FORCE REQUIREMENT

An object moving in a circle is experiencing an acceleration. Even if moving around the perimeter of the circle with a constant speed, there is still a change in velocity and subsequently an acceleration. This acceleration is directed towards the centre of the circle. And in accord with Newton's second law of motion, an object which experiences an acceleration must also be experiencing a net force. The direction of the net force is in the same direction as the acceleration. So for an object moving in a circle, there must be an inward force acting upon it in order to cause its inward acceleration. This is sometimes referred to as the centripetal force requirement. The word *centripetal* (not to be confused with the F-word *centrifugal*) means centre-seeking. For object's moving in circular motion, there is a net force acting towards the centre which causes the object to *seek* the centre.

To understand the importance of a centripetal force, it is important to have a sturdy understanding of the Newton's first law of motion - the law of inertia. The law of inertia states that...

... objects in motion tend to stay in motion with the same speed and the same direction unless acted upon by an unbalanced force.

According to Newton's first law of motion, it is the natural tendency of all moving objects to continue in motion in the same direction that they are moving... unless some form of unbalanced force acts upon the object to deviate its motion from its straight-line path. Moving objects will tend to naturally travel in straight lines; an unbalanced force is only required to cause it to turn. Thus, the presence of an unbalanced force is required for objects to move in circles.

INERTIA, FORCE AND ACCELERATION FOR AN AUTOMOBILE PASSENGER

The idea expressed by Newton's law of inertia should not be surprising to us. We experience this phenomenon of inertia nearly everyday when we drive our automobile. For example, imagine that you are a passenger in a car at a traffic light. The light turns green and the driver accelerates from rest. The car begins to accelerate forward, yet relative to the seat which you are on your body begins to lean backwards. Your body being at rest tends to stay at rest. This is one aspect of the law of inertia - "objects at rest tend to stay at rest." As the wheels of the car spin to generate a forward force upon the car and cause a forward acceleration, your body tends to stay in place. It certainly might seem to you as though your body were experiencing a backwards force causing it to accelerate backwards. Yet you would have a difficult time identifying such a backwards force on your body. Indeed there isn't one. The feeling of being thrown backwards is merely the tendency of your body to resist the acceleration and to remain in its state of rest. The car is accelerating out from under your body, leaving you with the false feeling of being pushed backwards.

Now imagine that you are in the same car moving along at a constant speed approaching a stoplight. The driver applies the brakes, the wheels of the car lock, and the car begins to skid to a stop. There is a backwards force upon the forward moving car and subsequently a backwards acceleration on the car. However, your body, being in motion, tends to continue in motion while the car is skidding to a stop. It certainly might seem to you as though your body were experiencing a forwards force causing it to accelerate forwards. Yet you would once more have a difficult time identifying such a forwards force on your body. Indeed there is no physical object accelerating you forwards. The feeling of being thrown forwards is merely the tendency of your body to resist the deceleration and to remain in its state of forward motion. This is the second aspect of Newton's law of inertia - "an object in motion tends to stay in motion with the same speed and in the same direction...." The unbalanced force acting upon the car causes the car to slow down while your body continues in its forward motion. You are once more left with the false feeling of being pushed in a direction which is opposite your acceleration.

These two driving scenarios are summarized by the following graphic.

Car Starts from Rest

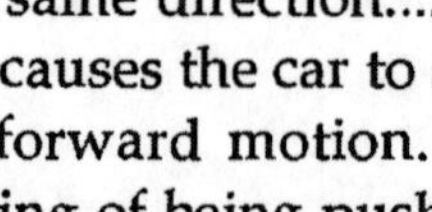

A passenger at rest would remain at rest, thus causing the "sensation of a backwards accelaration."

Moving Car Brakes to Stop

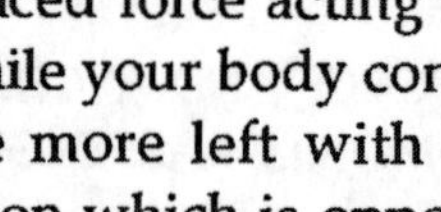

A passenger in motion would remain in motion, thus causing the "sensation of a forwards acceleration."

In each case - the car starting from rest and the moving car braking to a stop - the direction which the passengers lean is opposite the direction of the acceleration. This is merely the result of the passenger's inertia - the tendency to resist acceleration. The passenger's lean is not an acceleration in itself but rather the tendency to maintain the state of motion while the car does the acceleration. The tendency of a passenger's body to maintain its state of rest or motion while the surroundings (the car) accelerate is often misconstrued as an acceleration. This becomes particularly problematic when we consider the third possible inertia experience of a passenger in a moving automobile - the left hand turn.

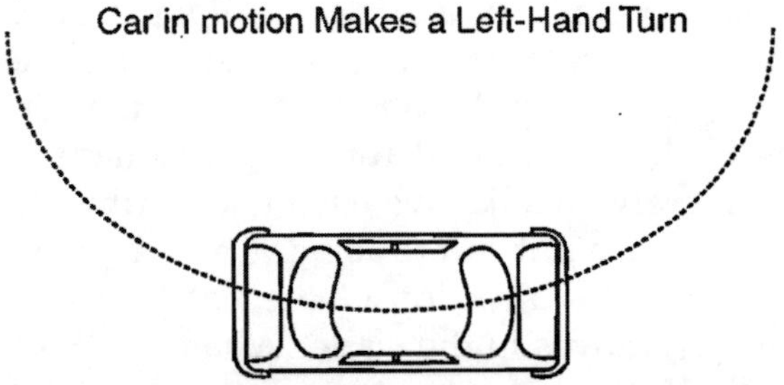

A passenger in motion would remain in motion with the same speed and in the same direction, thus causing the "sensation of an outwards acceleration."

Suppose that on the next part of your travels the driver of the car makes a sharp turn to the left at constant speed. During the turn, the car travels in a circular-type path. That is, the car sweeps out one-quarter of a circle. The friction force acting upon the turned wheels of the car cause an unbalanced force upon the car and a subsequent acceleration. The unbalanced force and the acceleration are both directed towards the centre of the circle about which the car is turning. Your body however is in motion and tends to stay in motion. It is the inertia of your body - the tendency to resist acceleration - which causes it to continue in its forward motion. While the car is accelerating inward, you continue in a straight line. If you are sitting on the passenger side of the car, then eventually the outside door of the car will hit you as the car turns inward. This phenomenon might cause you to think that you are being accelerated outwards away from the centre of the circle. In reality, you are continuing in your straight-line inertial path tangent to the circle while the car is accelerating out from under you. The sensation of an outward force and an outward acceleration is a false sensation. There is no physical object capable of pushing you outwards. You are merely experiencing the tendency of your body to continue in its path tangent to the circular path along which the car is turning. You are once more left with the false feeling of being pushed in a direction which is opposite your acceleration.

THE CENTRIPETAL FORCE AND DIRECTION CHANGE

Any object moving in a circle (or along a circular path) experiences a centripetal force. That is, there is some physical force pushing or pulling the object towards the centre of the circle. This is the centripetal force requirement. The word *centripetal* is merely an adjective used to describe the direction of the force. We are not introducing a new *type* of force but rather describing the direction of the net force acting upon the object which moves in the circle. Whatever the object, if it moves in a circle, there is some force acting upon it to cause it to deviate from its straight-line path, accelerate inwards and move along a circular path. Three such examples of centripetal force are shown below.

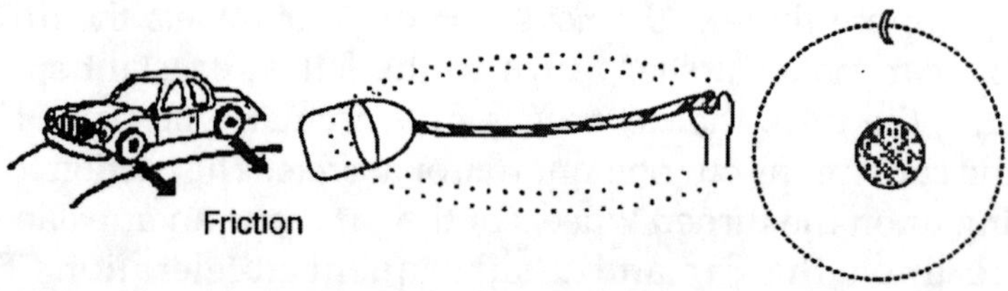

Fig. (i) As a car makes a turn, the force of friction acting upon the turned wheels of the car provides centripetal force required for circular motion. (ii) As a bucket of water is tied to a string and spun in a circle, the tension force acting upon the bucket provides the centripetal force required for circular motion. (iii) As the moon orbits the Earth, the force of gravity acting upon the moon provides the centripetal force required for circular motion.

The centripetal force for uniform circular motion alters the direction of the object without altering its speed. The idea that an unbalanced force can change the direction of the velocity vector but not its magnitude may seem a bit strange. How could that be? There are a number of ways to approach this question. One approach involves to analyse the motion from a work-energy standpoint. The work is a force acting upon an object to cause a displacement. The amount of work done upon an object is found using the equation

Work = Force * displacement * cosine (Theta)

where the Theta in the equation represents the angle between the force and the displacement. As the centripetal force acts upon an object moving in a circle at constant speed, the force always acts inward as the velocity of the object is directed tangent to the circle. This would mean that the force is always directed perpendicular to the direction which the object is being displaced. The angle Theta in the above equation is 90 degrees and the cosine of 90 degrees is 0. Thus, the work done by the centripetal force in the case of uniform circular motion is 0 Joules. When no work is done upon an object by external forces, the total mechanical energy (potential energy plus kinetic energy) of the object remains constant. So if an object is moving in a horizontal circle at constant speed, the centripetal force does not do work and cannot alter the total mechanical energy of the object. For this reason, the kinetic energy and therefore, the speed of the object will remain constant. The force can indeed accelerate the object - by changing its direction - but it cannot change its speed. In fact, whenever the unbalanced centripetal force acts perpendicular to the direction of motion, the speed of the object will remain constant. For an unbalanced force to change the speed of the object, there would have to be a component of force in the direction of (or the opposite direction of) the motion of the object.

A second approach to this question of why the centripetal force causes a direction change but not a speed change involves vector components and Newton's second law. The following imaginary scenario will be used to help illustrate the point.

THE FORBIDDEN F-WORD

When the subject of circular motion is discussed, it is not uncommon to hear mention of the word *centrifugal*. Centrifugal, not to be confused with centripetal, means away from the centre or outward. The use of or at least the familiarity with this word centrifugal, combined with the common sensation of an outward lean when experiencing circular motion, often creates or reinforces a common student misconception. The common misconception, believed by many physics students, is the notion that objects in circular motion are experiencing an outward force. "After all," a well-meaning student may think, Therefore, circular motion must be characterized by an outward force." This misconception is often fervently adhered to despite the clear presentation by a textbook or teacher of an inward force requirement. The motion of an object in a circle requires that there be an inward net force - the centripetal force requirement. There is an inward-directed acceleration which demands an inward force. Without this inward force, an object would maintain a straight-line motion tangent to the perimeter of the circle. Without this inward or centripetal force, circular motion would be impossible.

So why then is this student misconception of an outward or centrifugal force so prevalent and so stubbornly adhered to? Perhaps like all misconceptions, the notion of a centrifugal force as lodged in a person's head has a particularly lengthy history. Part of that history is certainly attributable to the experience of a circular motion - either as a passenger or driver in an automobile or perhaps on an amusement park ride. Even learned physics types would admit that circular motion leaves the moving person with the sensation of being thrown outward from the centre of the circle. But before drawing hasty conclusions, ask yourself three probing questions:

- Does the sensation of being thrown outward from the centre of a circle mean that there was definitely an outward force?
- If there is such an outward force on my body as I make a left-hand turn in an automobile, then what physical object is supplying the outward push or pull?

- And finally, could that sensation be explained in other ways which are more consistent with our growing understanding of Newton's laws?

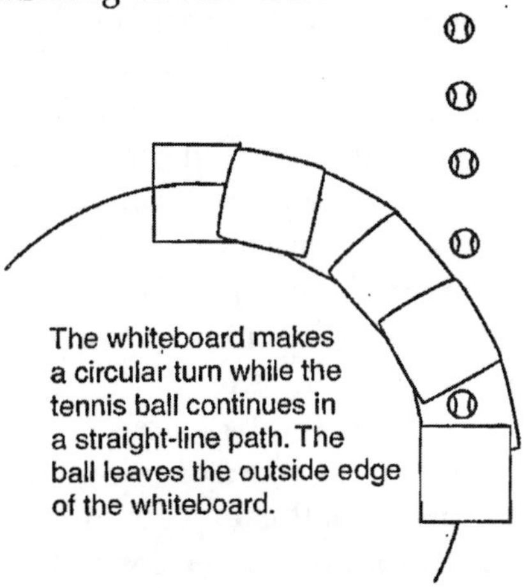

If you can answer the first of these questions with "No" then you have a chance. But if you quickly conclude that the outward feeling means there is an outward force, The sensation of being thrown outward is attributable to the idea of inertia, rather than the idea of force. When making that left-hand turn in the car, your tendency to be thrown rightward across the seat (which would be outward or away from the centre of the circle) was not due to a force. It was due to your tendency to travel in a straight line while the car seat was making its turn. In fact, you were not thrown rightward at all; you moved in a perfectly straight line. If an airborne camera had collected the motion on film from above and we could watch the instant replay, then it would be a no-brainer - the car turned left and your body kept going straight. Finally, your body hits the door on the right side of the car and the door provides an inward push on your body to cause your body to begin moving in circular motion. But until hitting the door, your body's tendency was to follow its inertial path.

A common physics demonstration involves using a flat whiteboard with a tennis ball on top of it. The whiteboard is carried along in a straight line path; the ball rest on top of the whiteboard and follows the same straight-line path. Then suddenly, the board is turned leftward to begin a circular motion; yet the ball keeps moving straight. Ultimately, the ball rolls off the right-edge of the board and continues in its straight-line inertial path. Without an unbalanced force on the ball, the ball continues in its original motion. The whiteboard merely moved out from under the ball as it makes its turn. If you could watch carefully, then you could view the ball's path from the perspective of an airborne camera. It's a no brainer - the ball moves straight while the whiteboard turns. And finally, the ball travels off the "outside edge" of the whiteboard. Relative to the circular motion of the whiteboard, the ball moves away from the centre of the circle. But explaining the motion of the ball does not require that we imagine or dream up the existence of an outward or centrifugal force. The motion of the ball is explained by the tendency of an object in motion to continue in motion in the same direction. INERTIA!

Now suppose that a block is attached to the top of the whiteboard on the "outside" of the ball with such an orientation that it would apply an inward force upon the ball. When the whiteboard is turned, the block would turn as well and supply the centripetal force required to move the ball in a circle. Without the block, the ball would have moved along the straight-line path, moving to position 1 after say 0.1 seconds, then to position 2 after 0.2 seconds, then to position 3 after 0.3 seconds, and so on. But with the block supplying an inward force, the ball moves inward towards the centre of the circle relative to its straight-line path. Instead of being at position 1, the ball is closer to the centre at position 1'. And instead of being at position 2 after 0.2 seconds, the ball is forced inwards towards position 2'. And instead of being at position 3 after 0.3 seconds, the ball is forced inwards towards position 3'. The inward net force accelerates the ball inward, causing it to deviate from its straight-line path that is directed tangent to the circle.

If you were the tennis ball in the first example above, then you might feel like you were being pushed outwards. After all, you would travel through the *outside door of the whiteboard*. Yet it is clear from the diagram and the discussion, that you are not deviating from any straight-line path. It is merely that the whiteboard is moving inward relative to your path and you are moving outward relative to the whiteboard's path. But this sensation of relative motion does not give reason for supposing that an outward force exists. This notion of an outward force is merely fictitious. Newton's law of inertia - "an object in motion continues in motion with the same speed and in the same direction unless acted upon an unbalanced force" - provides a more reasonable explanation for the sensations experienced by those who are in circular motion. A centrifugal or outward net force simply does not exist. No physical object could ever be identified that was pushing you outwards. And if there was a physical object pushing or pulling you outwards (e.g., in the rightwards direction when taking a left-hand turn), then you certainly would not turn in the circle that you are turning in.

An object moving n circular motion is at all times moving tangent to the circle; the velocity vector for the object is directed tangentially. To make the circular motion, there must be a net or unbalanced force directed towards the centre of the circle in order to deviate the object from its otherwise tangential path. This path is an inward force - a centripetal force. That is spelled c-e-n-t-r-i-p-e-t-a-l, with a "p." The other word - centrifugal, with an "f" - will be considered our forbidden F-word. Simply don't use it and please don't believe in it.

MATHEMATICS OF CIRCULAR MOTION

There are three mathematical quantities which will be of primary interest to us as we analyse the motion of objects in circles. These three quantities are speed, acceleration and force. The speed of an object moving in a circle is given by the following equation.

$$\text{Average Speed} = \frac{\text{distance}}{\text{time}} = \frac{2 * p * R}{T}$$

where R represents radius, T represents period

The acceleration of an object moving in a circle can be determined by either two of the following equations.

$$\text{Acceleration} = \frac{v^2}{R}$$

$$\text{Acceleration} = \frac{4 * \pi^2 * R}{T^2}$$

where v represents speed, R represents radius, T represents period

The equation on the right (above) is derived from the equation on the left by the substitution of the expression for speed.

The net force (F_{net}) acting upon an object moving in circular motion is directed inwards. While there may by more than one force acting upon the object, the vector sum of all of them should add up to the net force. In general, the inward force is larger than the outward force (if any) such that the outward force cancels and the unbalanced force is in the direction of the centre of the circle. The net force is related to the acceleration of the object (as is always the case) and is thus given by the following three equations:

$$F_{net} = m * a \qquad F_{net} = m * \frac{v^2}{R} \qquad F_{net} = m * \frac{4 * \pi^2 * R}{T^2}$$

where m represents mass, v represents speed, R represents radius, T represents period

The equations in the middle (above) and on the right (above) are derived from the equation on the left by the substitution of the expressions for acceleration.

This set of circular motion equations can be used in two ways:

- As a "recipe" for algebraic problem-solving in order to solve for an unknown quantity.
- As a *guide to thinking* about how an alteration in one quantity would affect a second quantity.

Equations as a Guide to Thinking

An equation expresses a mathematical relationship between the quantities present in that equation. For instance, the equation for Newton's second law identifies how acceleration is related to the net force and the mass of an object.

$$a = \frac{F_{net}}{m}$$

The relationship expressed by the equation is that the acceleration of an object is directly proportional to the net force acting upon it. In other words, the bigger the net force value is, the bigger that the acceleration value will be. As net force increases, the acceleration increases. In fact, if the net force were increased by a factor of 2, the equation would predict that the acceleration would increase by a factor of 2. Similarly, if the net force were decreased by a factor of 2, the equation would predict that the acceleration would decrease by a factor of 2.

Newton's second law equation also reveals the relationship between acceleration and mass. According to the equation, the acceleration of an object is inversely proportional to mass of the object. In other words, the bigger the mass value is, the smaller that the acceleration value will be. As mass increases, the acceleration decreases. In fact, if the mass were increased by a factor of 2, the equation would predict that the acceleration would decrease by a factor of 2. Similarly, if the mass were decreased by a factor of 2, the equation would predict that the acceleration would increase by a factor of 2.

As mentioned previously, equations allow for predictions to be made about the affect of an alteration of one quantity on a second quantity. Since the Newton's second law equation shows three quantities, each raised to the first power, the predictive ability of the equation is rather straightforward. The predictive ability of an equation becomes more complicated when one of the quantities included in the equation is raised to a power. For instance, consider the following equation relating the net force (F_{net}) to the speed (v) of an object moving in uniform circular motion.

$$F_{net} = m * \frac{v^2}{R}$$

This equation shows that the net force required for an object to move in a circle is directly proportional to the square of the speed of the object. For a constant mass and radius, the F_{net} is proportional to the speed^2.

$$F_{net} = m * \frac{v^2}{R}$$

↓ for a constant mass(m) and constant radius(R)

$$F_{net} \sim v^2$$

The factor by which the net force is altered is the square of the factor by which the speed is altered. Subsequently, if the speed of the object is doubled, the net force required for that object's circular motion is quadrupled. And if the speed of the object is halved, the net force required is decreased by a factor of 4.

Equations as a Recipe for Problem-Solving

The mathematical equations presented above for the motion of objects in circles can be used to solve circular motion problems in which an unknown quantity must be determined. The process of solving a circular motion problem is much like any other problem in physics class. The process involves a careful reading of the problem, the identification of the known and required information in variable form, the selection of the relevant equation(s), substitution of known values into the equation, and finally algebraic manipulation of the equation to determine the answer. Consider the application of this process to the following two circular motion problems.

APPLICATIONS OF CIRCULAR MOTION

Roller Coaster G-Forces

Roller coaster rides are notorious for creating accelerations and *g-forces* which are capable of transforming stomach contents into airborne projectiles. As a rider starts the descent down the first drop, she begins a one-minute adventure filled with various sensations of weightlessness, heaviness, and *jerkiness*. The parts of the ride which are most responsible for these sensations of weightlessness and heaviness are the *clothoid loops*.

The explanation for the various sensations experienced on a roller coaster loop are associated with Newton's laws of motion and the physics of circular motion.

A clothoid loop has a constantly curving shape with sections which resemble the curve of a circle (in actuality, it is considered to be a section of a *cornu spiral* having a constantly changing radius). A coaster rider is continuously altering her direction of motion while moving through the loop. At all times, the direction of motion could be described as being tangent to the loop.

This change in direction is caused by the presence of unbalanced forces and results in an acceleration. Not only is there an acceleration, the magnitude and direction of the acceleration is continuously changing. Within nearly a one second time interval, the riders may experience accelerations of 20 m/s/s downwards to 30 m/s/s upwards; such drastic changes in acceleration normally occur as the rider moves from the top of the loop to the bottom of the loop. These drastic changes in accelerations are the cause of much of the thrill experienced by coaster riders.

To understand the feelings of weightlessness and heaviness experienced while riding through a loop, it is important to think about the forces acting upon the riders. To simplify the discussion, we will assume that there are negligible amounts of air resistance acting upon the riders.

Thus, the only forces exerted upon the riders are the force of gravity and the normal force (the force of the seat pushing up on the rider). The force of gravity is at all times directed downwards and the normal force is at all times directed perpendicular to the seat of the car. Since the orientation of the car on the track is continuously changing, the normal force is continuously changing its direction. The magnitude and direction of these two forces during the motion through the loop are depicted in the animation below.

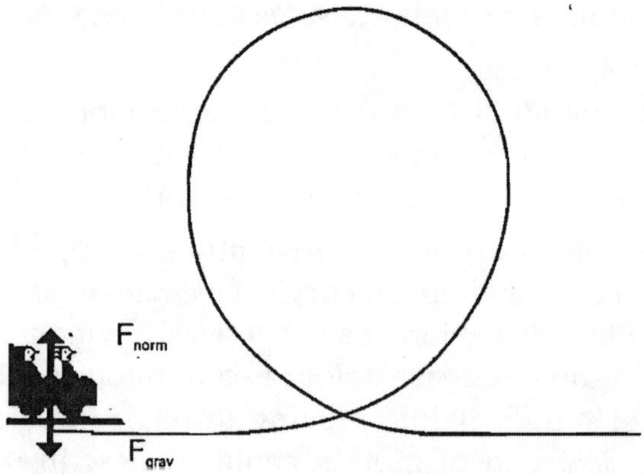

For an object to move along a circular path at a constant speed, there must be a net *inward force* acting upon the rider. This is commonly referred to as the centripetal force requirement. The motion through a coaster loop isn't precisely an example of moving in a circle at constant speed since the loop is neither circular not the speed constant. Nonetheless, because of the similarity of the motion along the loop's path to uniform circular motion, principles of uniform circular motion can be applied to the rider.

The net force acting upon the rider has an *inwards* direction (towards the centre of the circle). Since the net force is the vector sum of all the forces, the head-to-tail addition of the normal force and the gravity force should sum to a resultant force which points inward. The diagram below depicts the free-body diagrams for a rider at four locations along the loop. The diagram also shows that the vector sum of the two forces (i.e., the net force) points mostly towards the centre of the loop for each of the locations.

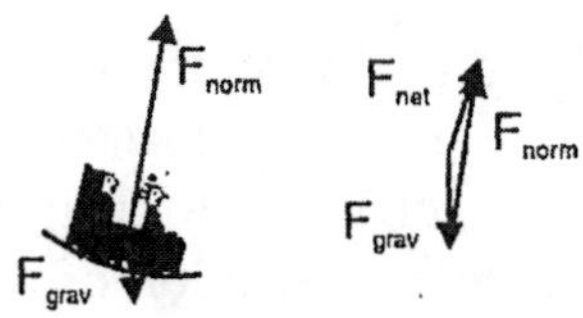

Botton of Loop

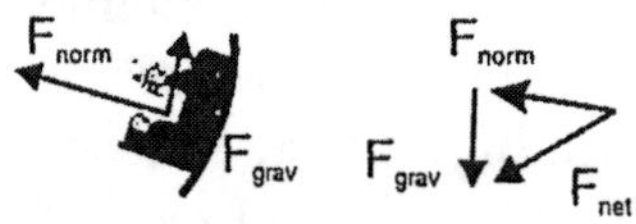

Right Side of Loop

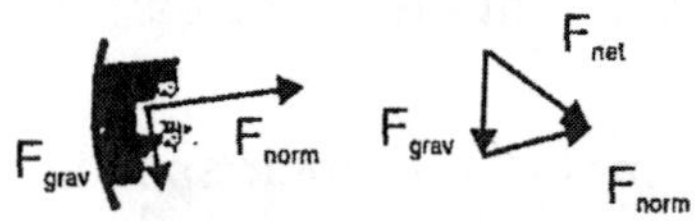

Left Side of Loop

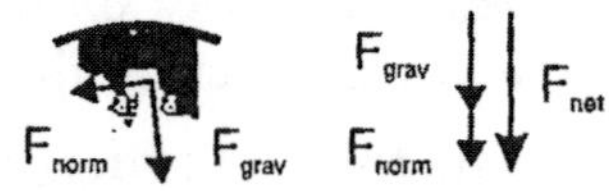

Top of Loop

Feelings of weightlessness and heaviness are associated with the normal force; they have little to do with the force of gravity. A person who feels weightless has not lost weight. The force of gravity acting upon the person is the same magnitude as it always is. Observe that in the animation above the force of gravity is everywhere the same. The normal force however has a small magnitude at the top of the loop (where the rider often feels weightless) and a large magnitude at the bottom of the loop (where the rider often feels heavy). The normal force is large at the bottom of the loop because in order for the net force to be directed inward, the normal force must be greater than the *outward* gravity force. At the top of the loop, the gravity force is directed inward and thus, there is no need for a large normal force in order to sustain the *circular* motion. The fact that a rider experiences a large force exerted by the seat upon her body when at the bottom of the loop is the explanation of why she *feels* heavy. In actuality, she is not heavier; she is only experiencing the large magnitude of force which is normally exerted by seats upon heavy people while at rest.

Roller Coasters and Amusement Park Physics

Americans are wild about amusement parks. Each day, we flock by the millions to the nearest park, paying a sizable hunk of money to wait in long lines for a short 60-second ride on our favourite roller coaster. The thought prompts one to consider what is it about a roller coaster ride that provides such widespread excitement among so many of us and such dreadful fear in the rest? Is our excitement about coasters due to their high speeds? Absolutely not! In fact, it would be foolish to spend so much time and money to ride a selection of roller coasters if it were for reasons of speed. It is more than likely that most of us sustain higher speeds on our ride along the interstate highway on the way to the amusement park than we do once we enter the park.

The thrill of roller coasters is not due to their speed, but rather due to their accelerations and to the feelings of weightlessness and weightiness which they produce. Roller coasters thrill us because of their ability to accelerate us downward one moment and upwards the next; leftwards one moment and rightwards the next. Roller coasters are about acceleration; that's what makes them thrilling. We will focus on the centripetal acceleration experienced by riders within the circular-shaped sections of a roller coaster track. These sections include the clothoid loops (which we will approximate as a circle), the sharp 180-degree banked turns, and the small dips and hills found along otherwise straight sections of the track.

Circular Motion Sections on Roller Coaster Rides

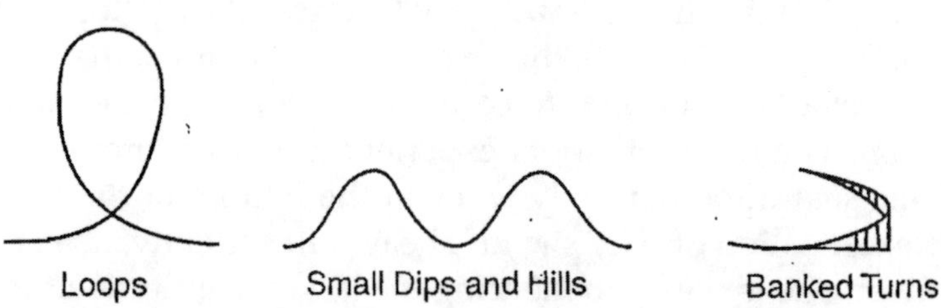

The most obvious section on a roller coaster where centripetal acceleration occurs is within the so-called clothoid loops. Roller coaster loops assume a tear-dropped shape which is geometrically referred to as a clothoid. A clothoid is a section of a spiral in which the radius is constantly changing. Unlike a circular loop in which the radius is a constant value, the radius at the bottom of a clothoid loop is much larger than the radius at the top of the clothoid loop. A mere inspection of a clothoid reveals that the amount of curvature at the bottom of the loop is less than the amount of curvature at the top of the loop. To simplify our analysis of the physics of clothoid loops, we will approximate a clothoid loop as being a series of overlapping or adjoining circular sections. The radius of these circular sections is decreasing as one approaches the top of the loop.

Furthermore, we will limit our analysis to two points on the clothoid loop - the top of the loop and the bottom of the loop. For this reason, our analysis will focus on the two circles which can be matched to the curvature of these two sections of the clothoid. The diagram at the right shows a clothoid loop with two circles of different radius inscribed into the top and the bottom of the loop. Note that the radius at the bottom of the loop is significantly larger than the radius at the top of the loop.

The Clothoid Loop

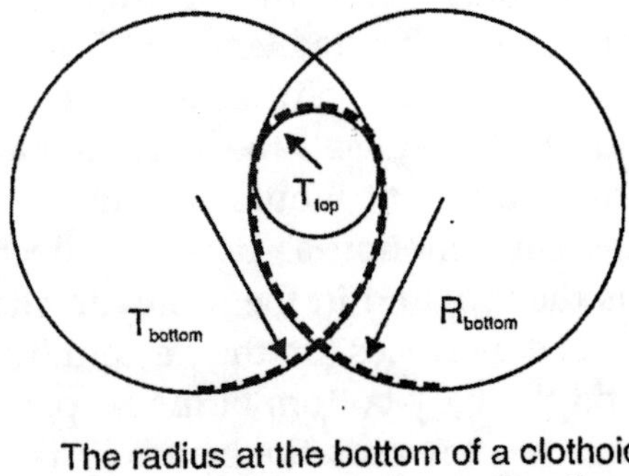

The radius at the bottom of a clothoid loop is significantly larger than the radius at the top of the clothoid loop.

As a roller coaster rider travels through a clothoid loop, she experiences an acceleration due to both a change in speed and a change in direction. A rightward moving rider gradually becomes an upward moving rider, then a leftward moving rider, then a downward moving rider, before finally becoming a rightward-moving rider once again. There is a continuous change in the direction of the rider as she moves through the clothoid loop. A change in direction is one characteristic of an accelerating object. In addition to changing directions, the rider also changes speed. As the rider begins to ascend (climb upward) the loop, she begins to slow down. As energy principles would suggest, an increase in height (and in turn an increase in potential energy) results in a decrease in kinetic energy and speed. And conversely, a decrease in height (and in turn a decrease in potential energy) results in an increase in kinetic energy and speed. So the rider experiences the greatest speeds at the bottom of the loop - both upon entering and leaving the loop - and the lowest speeds at the top of the loop.

This change in speed as the rider moves through the loop is the second aspect of the acceleration which a rider experiences. For a rider moving through a circular loop with a constant speed, the acceleration can be described as being centripetal or towards the centre of the circle. In the case of a rider moving through a noncircular loop at non-constant speed, the acceleration of the rider has two components. There is a component which is directed towards the centre of the circle (a_c) and attributes itself to the direction change; and there is a component which is directed tangent (a_t) to the track (either in the opposite or in the same direction as the car's direction of motion) and attributes itself to the car's change in speed. This tangential component would be directed opposite the direction of the car's motion as its speed decreases (on the ascent towards the top) and in the same direction as the car's motion as its speed increases (on the descent from the top). At the very top and the very bottom of the loop, the acceleration is primarily directed towards the centre of the circle. At the top, this would be in the downward direction and at the bottom of the loop it would be in the upward direction.

The inwards acceleration of an object is caused by an inwards net force. Circular motion (or merely motion along a curved path) requires an inwards component of net force. If all the forces which act upon the object are added together as vectors, then the net force would be directed inwards. Neglecting friction and air resistance, a roller coaster car will experience two forces: the force of gravity (F_{grav}) and the normal force (F_{norm}). The normal force is directed in a direction perpendicular to the track and the gravitational force is always directed downwards. We will concern ourselves with the relative magnitude and direction of these two forces for the top and the bottom of the loop. At the bottom of the loop, the track pushes upwards upon the car with a normal force. However, at the top of the loop the normal force is directed downwards; since the track (the supplier of the normal force) is above the car, it pushes downwards upon the car. The free-body diagram for these two positions are shown in the diagrams at the right.

The magnitude of the force of gravity acting upon the passenger (or car) can easily be found using the equation $F_{grav} = m \cdot g$ where g = acceleration of gravity (9.8 m/s^2). The magnitude of the normal force depends on two factors - the speed of the car, the radius of the loop and the mass of the rider. As depicted in the free body diagram, the magnitude of F_{norm} is always greater at the bottom of the loop than it is at the top. The normal force must always be of the appropriate size to combine with the F_{grav} in such a way to produce the required inward or centripetal net force. At the bottom of the loop, the F_{grav} points outwards away from the centre of the loop. The normal force must be sufficiently large to overcome this F_{grav} and supply some excess force to result in a net inward force. In a sense, F_{grav} and F_{norm} are in a tug-of-war; and F_{norm} must win by an amount equal to the net force. At the top of the loop, both F_{grav} and F_{norm} are directed inwards. The F_{grav} is found in the usual way (using the equation $F_{grav} = m \cdot g$). Once more the F_{norm} must provide sufficient force to produce the required inward or centripetal net force.

The use of Newton's second law and free-body diagrams to solve circular motion diagrams was illustrated. It was emphasized at that time that any given physical situation can be analysed in terms of the individual forces which are acting upon an object. These individual forces must add up as vectors to the net force. Furthermore, the net force must be equal to the mass times the acceleration. The process of conducting a force analysis of a physical situation was first introduced in. Now we will investigate the use of these fundamental principles in the analysis of situations involving the motion of objects in circles.

Suggested Method of Solving Circular Motion Problems

1. From the verbal description of the physical situation, construct a free-body diagram. Represent each force by a vector arrow and label the forces according to type.
2. Identify the given and the unknown information (express in terms of variables such as m=, a=, v=, etc.).
3. If any of the individual forces are directed at angles to the horizontal and vertical, then use vector principles to resolve such forces into horizontal and vertical components.
4. Determine the magnitude of any known forces and label on the free-body diagram.

(For example, if the mass is given, then the F_{grav} can be determined. And as another example, if there is no vertical acceleration, then it is known that the vertical forces or force components balance, allowing for the possible determination of one or more of the individual forces in the vertical direction.)

5. Use circular motion equations to determine any unknown information.

(For example, if the speed and the radius are known, then the acceleration can be determined. And as another example, if the period and radius are known, then the acceleration can be determined.)

6. Use the remaining information to solve for the requested information.
   a. If the problem requests the value of an individual force, then use the kinematic information (R, T and v) to determine the acceleration and the F_{net}; then use the free-body diagram to solve for the individual force value.
   b. If the problem requests the value of the speed or radius, then use the values of the individual forces to determine the net force and acceleration; then use the acceleration to determine the value of the speed or radius.

Combine a force analysis with the above method to solve the following roller coaster problem.

Sample Roller Coaster Problem

Anna Litical is riding on The Demon at Great America. Anna experiences a downwards acceleration of 15.6 m/s^2 at the top of the loop and an upwards acceleration of 26.3 m/s^2 at the bottom of the loop. Use Newton's second law to determine the normal force acting upon Anna's 864 kg roller coaster car.

Steps 1 and 2 involve the construction of a free body diagram and the identification of known and unknown quantities. This is shown in below.

Free Body Diagrams

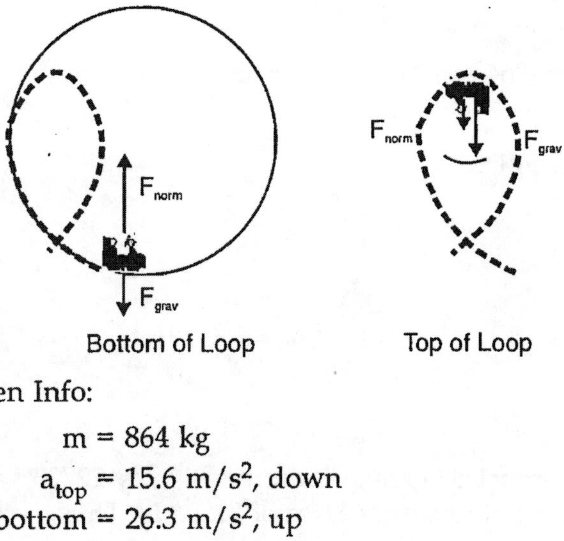

Bottom of Loop   Top of Loop

Given Info:

$$m = 864 \text{ kg}$$
$$a_{top} = 15.6 \text{ m/s}^2\text{, down}$$
$$\text{bottom} = 26.3 \text{ m/s}^2\text{, up}$$

Find:

F_{norm} at top and bottom

Step 3 of the suggested method would not apply to this problem since there are no forces directed "at angles" (that is, all the forces are either horizontally or vertically directed). Step 4 of the suggested method involves the determination of any known forces. In this case, the force of gravity can be determined from the equation $F_{grav} = m \bullet g$. Using a g value of 9.8 m/s², the force of gravity acting upon the 864-kg car is approximately 8467 N. Step 5 of the suggested method would be used if the acceleration were not given. In this instance, the acceleration is known. If the acceleration were not known, then it would have to be calculated from speed and radius information.

Step 6 of the suggested method involves the determination of an individual force - the normal force. This will involve a two-step process: first the net force (magnitude and direction) must be determined; then the net force must be used with the free body diagram to determine the normal force. This two-step process is shown below for the top and the bottom of the loop.

Bottom of Loop

$F_{net} = m * a$

$F_{net} = (864 \text{ kg}) * (26.3 \text{ m/s}^2\text{, up})$

$F_{net} = 22\,723 \text{ N, up}$

From FBD:

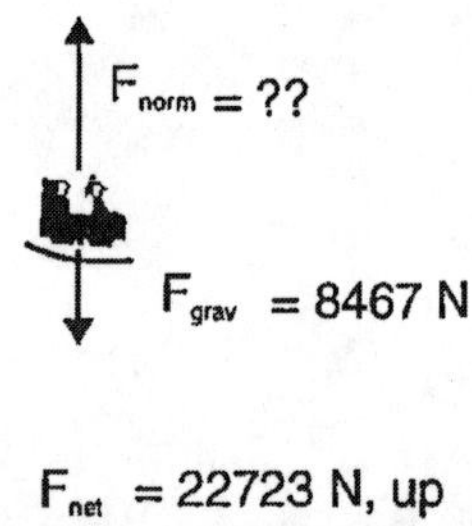

F_{norm} must be greater than the F_{grav} by 22723 N in order to supply a net upwards force of 22723 N. Thus,

$F_{norm} = F_{grav} + F_{net}$

$F_{norm} = 31190 \text{ N}$

Top of Loop

$F_{net} = m * a$

$F_{net} = (864 \text{ kg}) * (15.6 \text{ m/s}^2\text{, down})$

$F_{net} = 13478 \text{ N, down}$

From FBD:

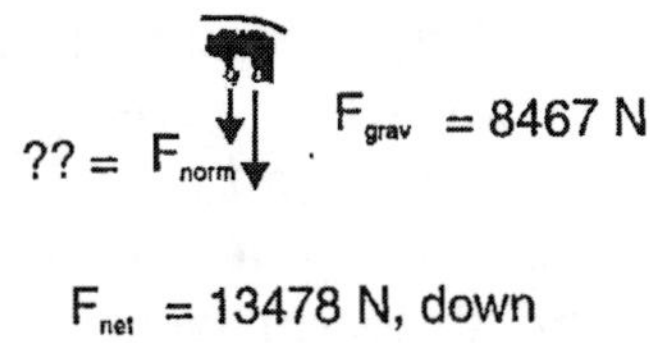

F_{norm} and F_{grav} together must combine together (i.e., add up) to supply the required inwards net force of 13478 N. Thus,

$F_{norm} = F_{net} - F_{grav}$

$F_{norm} = 5011$ N

Observe that the normal force is greater at the bottom of the loop than it is at the top of the loop. This becomes a reasonable fact when circular motion principles are considered. At all points along the loop - which we will refer to as circular in shape - there must be some inward component of net force. When at the top of the loop, the gravitational force is directed inwards (down) and so there is less of a need for a normal force in order to meet the net centripetal force requirement. When at the bottom of the loop, the gravitational force is directed outwards (down) and so now there is a need for a large upwards normal force in order to meet the centripetal force requirement. This principle is often demonstrated in a physics class using a bucket of water tied to a string. The water is spun in a vertical circle. As the water traces out its circular path, the tension in the string is continuously changing. The tension force in this demonstration is analogous to the normal force for a roller coaster rider. At the top of the vertical circle, the tension force is very small; and at the bottom of the vertical circle, the tension force is very large. (You might try this activity yourself outside with a small plastic bucket half-filled with water. Give extra caution to stay clear of all people, windows, trees and overhead power lines. Repeat enough cycles to observe the noticeable difference in tension force when the bucket is at the top and the bottom of the circle.)

If you have ever been on a roller coaster ride and traveled through a loop, then you have likely experienced this small normal force at the top of the loop and the large normal force at the bottom of the loop. The normal force provides a *feel* for a person's weight. The more you weigh, the more normal force which you will experience when at rest in your seat. But if you board a roller coaster ride and accelerate through circles (or clothoid loops), then you will feel a normal force which is constantly change and different from that which you are accustomed to.

This normal force provides a sensation or feeling of weightlessness or weightiness. When at the top of the loop, a rider will feel partially weightless if the normal forces become less than the person's weight. And at the bottom of the loop, a rider will feel very "weighty" due to the increased normal forces. It is important to realise that the force of gravity and the weight of your body are not changing. Only the magnitude of the supporting normal force is changing! There is some interesting history (and physics) behind the gradual usage of clothoid loops in roller coaster rides. In the early days of roller coaster loops, circular loops were used. There were a variety of problems, some of which resulted in fatalities, as the result of the use of these circular loops. Coaster cars entering circular loops at high speeds encountered excessive normal forces that were capable of causing whiplash and broken bones. Efforts to correct the problem by lowering entry speeds resulted in the inability of cars to make it around the entire loop without falling out of the loop when reaching the top.

The decrease in speeds as the cars ascended the large circular loop resulted in coaster cars turning into projectile cars (a situation known to be not good for business). The solution to the problem involved using low entry speeds and a loop with a sharper curvature at the top than at the bottom. Since clothoid loops have a continually changing radius, the radius is large at the bottom of the loop and shortened at the top of the loop. The result is that coaster cars can enter the loops at high speeds; yet due to the large radius, the normal forces do not exceed *3.5 G's*. At the top of the loop, the radius is small thus allowing a lower speed car to still maintain contact with the track and successfully make it through the loop. The clothoid loop is a testimony to an engineer's application of the centripetal acceleration equation - $a = v^2/R$. Now that's physics for better living!

The above discussion and force analysis applies to the circular-like motion of a roller coaster car in a clothoid loop. The second section along a roller coaster track where circular motion is experienced is along the small dips and hills. These sections of track are often found near the end of a roller coaster ride and involve a series of small hills followed by a sharp drop. Riders often feel heavy as they ascend the hill (along regions A and E in the diagram below). Then near the crest of the hill (regions B and F), their upward motion makes them feel as though they will fly out of the car; often times, it is only the safety belt which prevents such a mishap.

As the car begins to descend the sharp drop, riders are momentarily in a state of free fall (along regions C and G in the diagram below). And finally as they reach the bottom of the sharp dip (regions D and H), there is a large upwards force which slows their downwards motion. The cycle is often repeated mercilessly, churning the riders' stomachs and mixing the afternoon's cotton candy into a slurry of.... These small dips and hills combine the physics of circular motion with the physics of projectiles in order to produce the ultimate thrill of acceleration - rapidly changing magnitudes and directions of acceleration. The diagram below shows the various direction of accelerations which riders would experience along these hills and dips.

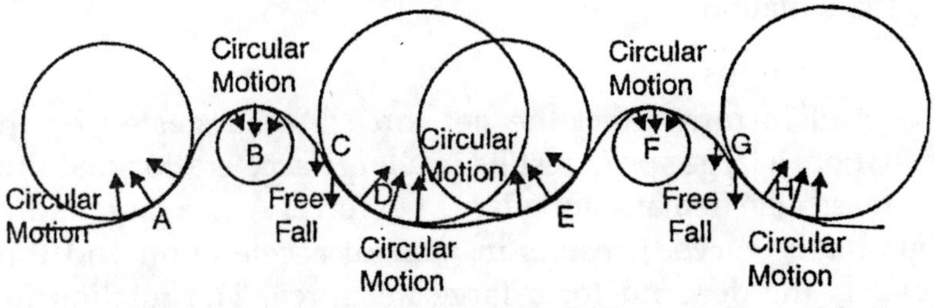

At various locations along these hills and dips, riders are momentarily traveling along a circular shaped arc. The arc is part of a circle - these circles have been inscribed on the above diagram in blue. In each of these regions there is an inward component of acceleration (as depicted by the black arrows). This inward acceleration demands that there also be a force directed towards the centre of the circle. In region A, the centripetal force is supplied by the track pushing normal to the track surface. Along region B, the centripetal force is supplied by the force of gravity and possibly even the safety mechanism/bar. At especially high speeds, a safety bar must supply even extra downward force in order to pull the riders downward and supply the remaining centripetal force required for circular motion. There are also wheels on the car which are usually tucked under the track and pulled downward by the track. Along region D, the centripetal force is once more supplied by the normal force of the track pushing upwards upon the car.

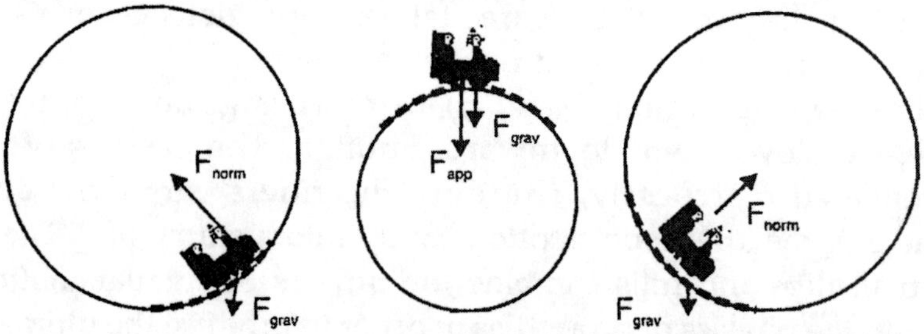

The magnitude of the normal forces along these various regions is dependent upon how sharply the track is curved along that region (the radius of the circle) and the speed of the car. These two variables affect the acceleration according to the equation

$$a = v^2 / R$$

and in turn affect the net force. As suggested by the equation, a large speed results in a large acceleration and thus increases the demand for a large net force. And a large radius (gradually curved) results in a small acceleration and thus lessens the demand for a large net force. The relationship between speed, radius, acceleration, mass and net force can be used to determine the magnitude of the *seat force* (i.e., normal force) upon a roller coaster rider at various sections of the track. The sample problem below illustrates these relationships. In the process of solving the problem, the same problem-solving strategy enumerated above will be utilized.

Sample Roller Coaster Problem

Anna Litical is riding on The American Eagle at Great America. Anna is moving at 18.9 m/s over the top of a hill which has a radius of curvature of 12.7 m. Use Newton's second law to determine the magnitude of the applied force of the track pulling down upon Anna's 621 kg roller coaster car.

Steps 1 and 2 involve the construction of a free body diagram and the identification of known and unknown quantities. This is shown in below.

Free Body Diagram

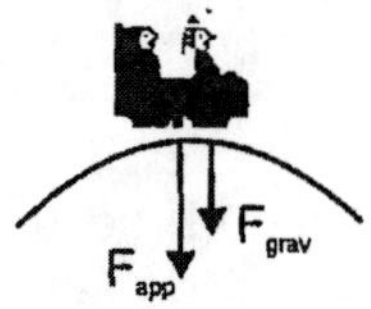

Given Info:

m = 621 kgv = 18.9 m/s

R = 12.7 m

Find:

F_{app} at top of hill

Step 3 of the suggested method would not apply to this problem since there are no forces directed "at angles" (that is, all the forces are either horizontally or vertically directed). Step 4 of the suggested method involves the determination of any known forces. In this case, the force of gravity can be determined from the equation $F_{grav} = m * g$. So the force of gravity acting upon the 621-kg car is approximately 6086 N. Step 5 of the suggested method involves the calculation of the acceleration from the given values of the speed and the radius. Using the equation, the acceleration can be calculated as follows

$a = v^2 / R$

$a = (18.9 \text{ m/s})^2 / (12.7 \text{ m})$

$a = 28.1 \text{ m/s}^2$

Step 6 of the suggested method involves the determination of an individual force - the applied force. This will involve a two-step process: first the net force (magnitude and direction) must be determined; then the net force must be used with the free body diagram to determine the applied force. This two-step process is shown below.

$F_{net} = m \bullet a$

$F_{net} = (621 \text{ kg}) \bullet (28.1 \text{ m/s}^2, \text{down})$

$F_{net} = 17467 \text{ N, down}$

As shown in FBD at right:

$F_{app} = F_{net} - F_{grav}$

$F_{norm} = 11381 \text{ N}$

F_{app} and F_{grav} must combine together (i.e., add up) to supply the required downwards net force of 17467 N.

This same method could be applied for any region of the track in which roller coaster riders momentarily experience circular motion.

Athletics

Circular motion is common to almost all sporting events. Whether it be sports car racing or track and field, baseball running or ice skating, the motion of objects in circles is a common observation of sports viewers around the world. Like any object moving in a circle, the motion of these objects which we view from the stadium bleachers or watch upon the television monitor are governed by Newton's laws of motion. Their circular motion - however brief or prolonged they may be - is characterized by an inward acceleration and caused by an inward net force. The mathematical analysis of such motions can be conducted in the manner presented. We will investigate a variety of applications of circular motion principles to the world of sports and use Newton's laws to mathematically analyse such motions. The emphasis will not be upon an investigation of the details of every possible sport; but rather upon learning how to apply some general principles so that they can subsequently be applied to every sport.

The most common example of the physics of circular motion in sports involves the turn. It could be a halfback in football making a turn around the corner of the line. Or it could be a softball player running the bases and making a turn around second base. Or it could be a bobsled in the Olympic games making a turn around a corner on the track. Whatever turning motion it happens to be, you can be sure that turning a corner involves circular motion principles. Now for certain not all turns involve a complete circle; nor do all turns have a perfectly circular shape. Some turns are only one-quarter of a turn - such as the fullback rounding the corner of the line in football. And some turns are hardly circular whatsoever. Nonetheless, any turn can be approximated as being a part of a larger circle or a part of several circles of varying size. A sharp turn can be considered part of a small circle. A more gradual turn is part of a larger circle. Some turns can begin sharply and gradually change in sharpness; or vice versa. In all cases, the motion around a turn can be approximated as part of a circle or a collection of circles. The diagram below depicts a variety of paths which a turn could make.

Any Turn Can be Considered Part of a Circle

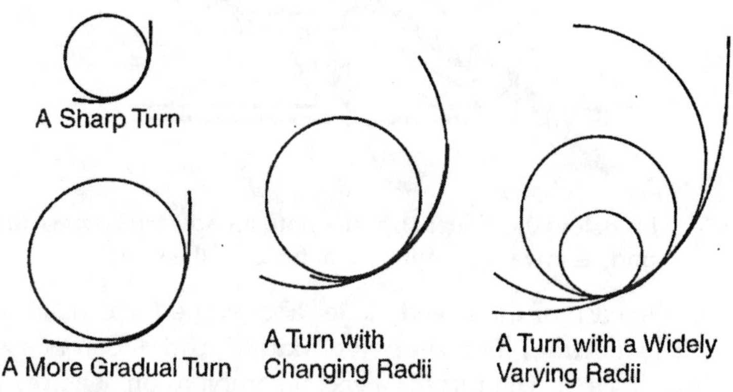

Because turning a corner involves the motion of an object which is momentarily moving along the path of a circle, both the concepts and the mathematics of circular motion can be applied to such a motion. Conceptually, such an object is moving with an inward acceleration - the inward direction being towards the centre of whatever *circle* the object is moving along. There would also be a centripetal force requirement for such a motion. That is, there must be some object supplying an inward force or inward component of force. When a person makes a turn on a horizontal surface, the person often *leans into the turn.* By leaning, the surface pushes upward at an angle *to the vertical.* As such, there is both a horizontal and a vertical component resulting from contact with the surface below. This contact force supplies two roles - it balances the downward force of gravity and meets the centripetal force requirement for an object in uniform circular motion. The upward component of the contact force is sufficient to balance the downward force of gravity and the horizontal component of the contact force pushes the person towards the centre of the circle. This contact force is depicted in the diagram below for a speed skater making a turn on ice.

Fig.The ice below the skater supplies both an upward component and an inward component of force on the skater.

In the case of the speed skater above, the force resulting from the contact between ice skates and ice has two components to it. The force is a vector combination of a normal force and a friction force. The normal force is the result of the stable surface providing support for any object pushing downward against it. The friction force is the result of the static friction force resulting from the ice-skate interaction. As the skater leans into the turn, she pushes downward and *outward* upon the ice. The high pressure and temperature of the blade upon the ice creates a shallow *groove* in which the blade momentarily rests.

The blade pushes outward upon the vertical wall of this groove and downward upon the floor of this groove. As we would expect from Newton's third law of motion, there is a *reaction force* of the ice pushing upward and inward upon the skate. If this blade-ice action does not occur, the skater could still lean and still try to push outward upon the ice. However, the blade would not get a *grip* upon the ice and the skater would be at risk of not making the turn. As a result, the ice skater's skates would move out from under her, she would fall to the ice, and she would travel in a straight-line inertial path. Without an inward force, the skater cannot travel through the turn.

The same principle of lean which allows the speed skater to make the turn around a portion of the circle applies to the wealth of other sporting events where participants lean into the turn in order to momentarily move in a circle. A downhill skier makes her turn by leaning into the snow. The snow pushes back in both an inward and an upward direction - balancing the force of gravity and supplying the centripetal force. A football player makes his turn by leaning into the ground. The ground pushes back in both an inward and upward direction - balancing the force of gravity and supplying both the centripetal force. A cyclist makes his turn in a similar manner as he leans at an angle to the horizontal. The road surface pushes with an upward component of force to balance the downward force of gravity. The road surface also pushes with a horizontal component of force towards the centre of the circle through which the cyclist is turning. A bobsled team makes their turn in a similar manner as they rise up onto the inclined section of track. Upon the incline, they naturally lean and the normal force acts at an angle to the vertical; this normal force supplies both the upward force to balance the force of gravity and the centripetal force to allow for the circular motion.

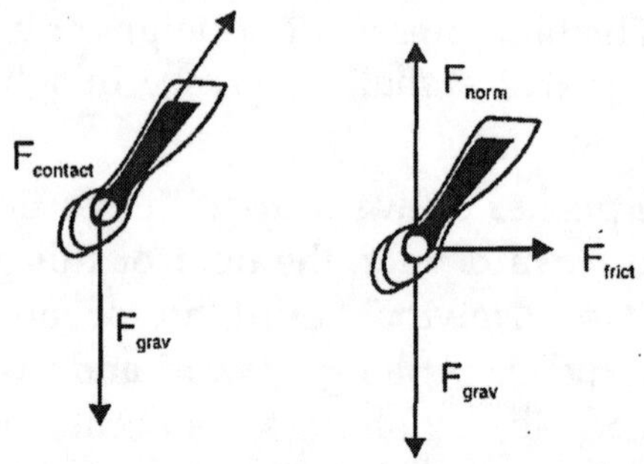

The contact force has two components-the vertical component balances the force of gravity and the horizontal component provides the inward force.

A turn is only possible when there is a component of force directed towards the centre of the circle about which the person is moving.

Mathematical Analysis of Turns in Athletics

The same mathematical equations which describe the motion of objects in circles applies to the motion of athletes making turns on the athletic field. It has been emphasized that any given physical situation can be analysed in terms of the individual forces which are acting upon an object; these individual forces must add up to the net force. Furthermore, the net force must be equal to the mass times the acceleration. The process of conducting a force analysis of a physical situation was first introduced. Now we will investigate the use of these fundamental principles in the analysis of situations involving the motion of athletes in circles. We will utilize the basic problem-solving approach. This approach can be summarized as follows.

Suggested Method of Solving Circular Motion Problems

1. From the verbal description of the physical situation, construct a free-body diagram. Represent each force by a vector arrow and label the forces according to type.
2. Identify the given and the unknown information (express in terms of variables such as m=, a=, v=, etc.).
3. If any of the individual forces are directed at angles to the horizontal and the vertical, then use vector principles to resolve such forces into horizontal and vertical components.
4. Determine the magnitude of any known forces and label on the free-body diagram.

   (For example, if the mass is given, then the F_{grav} can be determined. And as another example, if there is no vertical acceleration, then it is known that the vertical forces or force components balance, allowing for the possible determination of one or more of the individual forces in the vertical direction.)
5. Use circular motion equations to determine any unknown information.

   (For example, if the speed and the radius are known, then the acceleration can be determined. And as another example, if the period and radius are known, then the acceleration can be determined.)
6. Use the remaining information to solve for the requested information.
   a. If the problem requests the value of an individual force, then use the kinematic information (R, T and v) to determine the acceleration and the F_{net}; then use the free-body diagram to solve for the individual force value.
   b. If the problem requests the value of the speed or radius, then use the values of the individual forces to determine the net force and acceleration; then use the acceleration to determine the value of the speed or radius.

Combine a force analysis with the above method to solve the following circular motion problem.

Sample Speed Skater Problem

Bonnie is ice skating at the Olympic games. She is making a sharp turn with a radius of 22.6 m and with a speed of 16.1 m/s. Use Newton's second law to determine the acceleration and the angle of lean of Bonnie's 55.0-kg body.

Steps 1 and 2 involve the construction of a free body diagram and the identification of known and unknown quantities. This is shown in below.

Free Body Diagram

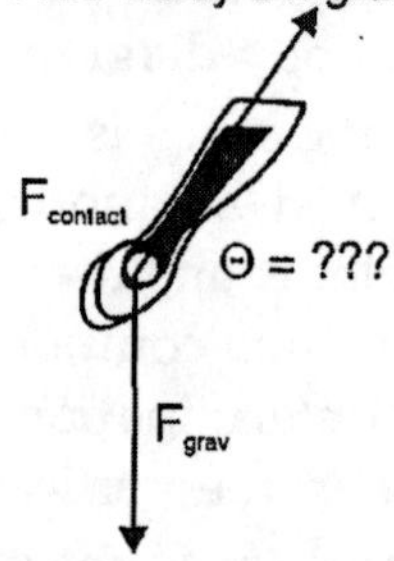

Given Info:

m = 55.0 kg

v = 16.1 m/s

r = 22.6 m

Find:

a = ???

Angle of lean = ???

Step 3 of the suggested method involves resolving any forces which act at angles into horizontal and vertical components. This is shown in the diagram at the right. The contact force can be broken into two components - F_{horiz} and F_{vert}. The vertical component of force would balance the force of gravity; and as such, the vertical component will be equal in magnitude to the force of gravity. The horizontal component of force remains unbalanced. As mentioned in the above discussion, this horizontal component is the net inward force;

and as such, F_{horiz} is equal to m*a. Finally, the two components are related to the angle of lean by the tangent function. Simple algebraic manipulation would yield the relationship shown in the graphic at the right. So the angle of lean can be found if the vertical and horizontal component of force are known.

Step 4 of the suggested method involves the determination of any known forces. In this case, the force of gravity can be determined from the equation $F_{grav} = m \bullet g$. o the force of gravity acting upon Bonnie's 55.0-kg body is approximately 539 N. And since this force is balanced by the vertical component of the contact force, the F_{vert} is also 539 N. Step 5 involves determination of Bonnie's acceleration as she makes the turn. This can be accomplished by using the acceleration equation for circular motion.

$$a = v^2/R$$

$$a = (16.1\ m/s)^2/(22.6\ m) = 11.5\ m/s^2$$

Now that the acceleration has been found, the angle of lean can be determined. As mentioned in the equation above, the angle of lean ("theta") can be determined from knowledge of the horizontal and vertical components of the contact force. The vertical component has already been calculated to be 539 N (equal to F_{grav}). And as previously mentioned, the horizontal component would be equal to F_{net}; this is shown below.

$$F_{horiz} = F_{net} = m \bullet a$$

$$F_{horiz} = (55.0\ kg) \bullet (11.5\ m/s/s) = 631\ N$$

Now finally the angle of lean can be determined.

$$\Theta = \text{invtan}\left(\frac{539N}{631N}\right) = 40.5\,\text{degrees}$$

The above problem illustrates the procedure of combining Newton's second law of motion with vector principles and circular motion equations in order to analyse turning motions of athletes. Now utilize the same general procedure described above to solve the following practice problem. When finished, click the button to view the answers.

Turning motions are not the only situations in sports in which people or objects move in circles. While turning motions are probably the most common examples of circular motion, they are not the only examples. There are certain track and field events - the hammer throw and the discus - in which athletes gather momentum in an object which is to be subsequently thrown. The pre-throw momentum is imparted to the projectile by whirling within a circle. Once momentum has been accumulated, the hammer or discus is launched into the air at an optimum angle in order to maximize the distance it travels. Regardless of the athletic event, the analysis of the circular motion remains the same. Newton's laws describe the force-mass-acceleration relationship; vector principles describe the relationship between individual forces and any angular forces; and circular motion equations describe the speed-radius-acceleration relationship.

UNIVERSAL GRAVITATION

Gravity is More Than a Name

Nearly every child knows of the word gravity. Gravity is the name associated with the mishaps of the milk spilled from the breakfast table to the kitchen floor and the youngster who topples to the pavement as the grand finale of the first bicycle ride. Gravity is the name associated with the reason for "what goes up, must come down," whether it be the baseball hit in the neighborhood sandlot game or the child happily jumping on the backyard mini-trampoline. We all know of the word gravity - it is the *thing* which causes objects to fall to Earth. Yet the role of physics is to do more than to associate words with phenomenon. The role of physics is to explain phenomenon in terms of underlying principles. The goal is to explain phenomenon in terms of principles which are so universal that they are capable of explaining more than a single phenomenon but a wealth of phenomenon in a consistent manner. Thus, a student's conception of gravity must grow in sophistication to the point that it becomes more than a mere name associated with falling phenomenon. Gravity must be understood in terms of its cause, its source, and its far-reaching implications on the structure and the motion of the objects in the universe.

Cannonballs shot from
"Newton's Mountain"

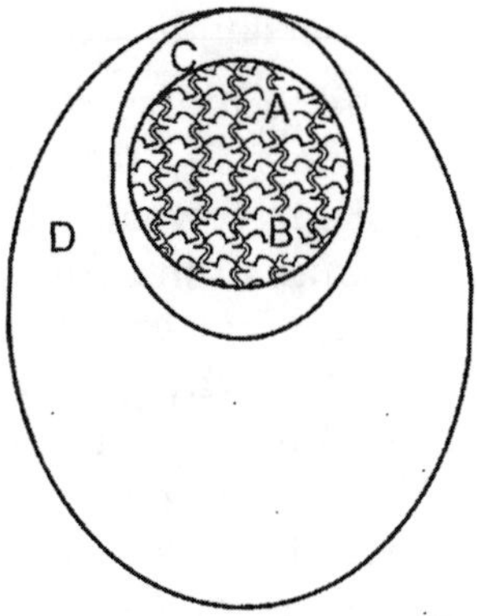

As launch speeds are increased, the connonballs travel greater distances before falling to earth. ultimately, the cannonballs will fall around the earth instead of into the earth.

Certainly gravity is a force which exists between the Earth and the objects which are near it. As you stand upon the Earth, you experience this force. We have become accustomed to calling it the force of gravity and have even represented it by the symbol F_{grav}. Most students of physics progress at least to this level of sophistication concerning the notion of gravity. This same force of gravity acts upon our bodies as we jump upwards from the Earth. As we rise upwards after our jump, the force of gravity slows us down. And as we fall back to Earth after reaching the peak of our motion, the force of gravity speeds us up. In this sense, the force gravity causes an acceleration of our bodies during this brief trip away from the earth's surface and back. In fact, many students of physics have become accustomed to referring to the actual acceleration of such an object as the acceleration of gravity. Not to be confused with the force of gravity (F_{grav}), the acceleration of gravity (g) is the acceleration experienced by an object when the only force acting upon it is the force of gravity. On and near Earth's surface, the value for the acceleration of gravity is approximately 9.8 m/s/s. It is the same acceleration value for all objects, regardless of their mass (and assuming that the only significant force is gravity). Many students of physics progress this far in their understanding of the notion of gravity.

We will build on this understanding of gravitation, making an attempt to understand the nature of this force. Many questions will be asked: How and by whom was gravity discovered? What is the cause of this force which we refer to with the name of gravity? What variables affect the actual value of the force of gravity? Why does the force of gravity acting upon an object depend upon the location of the object relative to the Earth? How does gravity affect objects which are far beyond the surface of the Earth? How far-reaching is gravity's influence? And is the force of gravity which attracts my body to the Earth related to the force of gravity between the planets and the Sun? These are the questions which will be pursued. And if you can successfully answer them, then the sophistication of your understanding has extended beyond the point of merely associating the name "gravity" with falling phenomenon.

THE APPLE, THE MOON, AND THE INVERSE SQUARE LAW

In the early 1600's, German mathematician and astronomer Johannes Kepler mathematically analysed known astronomical data in order to develop three laws to describe the motion of planets about the sun. Kepler's three laws emerged from the analysis of data carefully collected over a span of several years by his Danish predecessor and teacher, Tycho Brahe. Kepler's three laws of planetary motion can be briefly described as follows:

- The path of the planets about the sun are elliptical in shape, with the centre of the sun being located at one focus. (The Law of Ellipses)
- An imaginary line drawn from the centre of the sun to the centre of the planet will sweep out equal areas in equal intervals of time. (The Law of Equal Areas)
- The ratio of the squares of the periods of any two planets is equal to the ratio of the cubes of their average distances from the sun. (The Law of Harmonies)

$$\frac{g_{moon}}{g_{apple}} \frac{0.00272 m/s^2}{9.8 m/s^2} = \frac{1}{3600}$$

While Kepler's laws provided a suitable framework for describing the motion and paths of planets about the sun, there was no accepted explanation for why such paths existed. The cause for how the planets moved as they did was never stated. Kepler could only suggest that there was some sort of interaction between the sun and the planets which provided the driving force for the planet's motion. To Kepler, the planets were somehow "magnetically" driven by the sun to orbit in their elliptical trajectories. There was however no interaction between the planets themselves.

Newton was troubled by the lack of explanation for the planet's orbits. To Newton, there must be some cause for such elliptical motion. Even more troubling was the circular motion of the moon about the earth. Newton knew that there must be some sort of force which governed the heavens; for the motion of the moon in a circular path and of the planets in an elliptical path required that there be an inward component of force. Circular and elliptical motion were clearly departures from the inertial paths (straight-line) of objects. And as such, these celestial motions required a cause in the form of an unbalanced force. Circular motion (as well as elliptical motion) requires a centripetal force. The nature of such a force - its cause and its origin - bothered Newton for some time and was the fuel for much mental pondering. And according to legend, a breakthrough came at age 24 in an apple orchard in England. Newton never wrote of such an event, yet it is often claimed that the notion of gravity as the cause of all heavenly motion was instigated when he was struck in the head by an apple while lying under a tree in an orchard in England. Whether it is a myth or a reality, the fact is certain that it was Newton's ability to relate the cause for heavenly motion (the orbit of the moon about the earth) to the cause for Earthly motion (the falling of an apple to the Earth) which led him to his notion of universal gravitation.

A survey of Newton's writings reveals an illustration similar to the one shown at the right. The illustration was accompanied by an extensive discussion of the motion of the moon as a projectile. Newton's reasoning proceeded as follows. Suppose a cannonball is fired horizontally from a very high mountain in a region devoid of air resistance. In the absence of gravity, the cannonball would travel in a straight-line, tangential path. Yet in the presence of gravity, the cannonball would drop below this straight-line path and eventually fall to Earth (as in path A). Now suppose that the cannonball is fired horizontally again, yet with a greater speed. In this case, the cannonball would still fall below its straight-line tangential path and eventually drop to earth. Only this time, the cannonball would travel further before striking the ground (as in path B). Now suppose that there is a speed at which the cannonball could be fired such that the trajectory of the falling cannonball matched the curvature of the earth. If such a speed could be obtained, then the cannonball would fall around the earth instead of into it. The cannonball would fall towards the Earth without ever colliding into it and subsequently become a satellite orbiting in circular motion (as in path C). And then at even greater launch speeds, a cannonball would once more orbit the earth, but in an elliptical path (as in path D). The motion of the cannonball orbiting to the earth under the influence of gravity is analogous to the motion of the moon orbiting the Earth. And if the orbiting moon can be compared to the falling cannonball, it can even be compared to a falling apple. The same force which causes objects on Earth to fall to the earth also causes objects in the heavens to move along their circular and elliptical paths. Quite amazingly, the laws of mechanics which govern the motions of objects on Earth also govern the movement of objects in the heavens.

Of course, Newton's dilemma was to provide reasonable evidence for the extension of the force of gravity from earth to the heavens. The key to this extension demanded that he be able to show how the affect of gravity is diluted with distance. It was known at the time, that the force of gravity causes earthbound objects (such as falling apples) to accelerate towards the earth at a rate of 9.8 m/s^2. And it was also known that the moon accelerated towards the earth at a rate of 0.00272 m/s^2. If the same force which causes the acceleration of the apple to the earth also causes the acceleration of the moon towards the earth, then there must be a plausible explanation for why the acceleration of the moon is so much smaller than the acceleration of the apple. What is it about the force of gravity which causes the more distant moon to accelerate at a rate of acceleration which is approximately 1/3600-th the acceleration of the apple?

Newton knew that the force of gravity must somehow be "diluted" by distance. But how? What mathematical reality is intrinsic to the force of gravity which causes it to be inversely dependent upon the distance between the objects?

The riddle is solved by a comparison between the distance from the apple to the centre of the earth with the distance from the moon to the centre of the earth. The moon in its orbit about the earth is approximately 60 times further from the earth's centre than the apple is. The mathematical relationship becomes clear. The force of gravity between the earth and any object is inversely proportional to the square of the distance which separates that object from the earth's centre. The moon, being 60 times further away than the apple, experiences a force of gravity which is $1/(60)^2$ times that of the apple. The force of gravity follows an inverse square law.

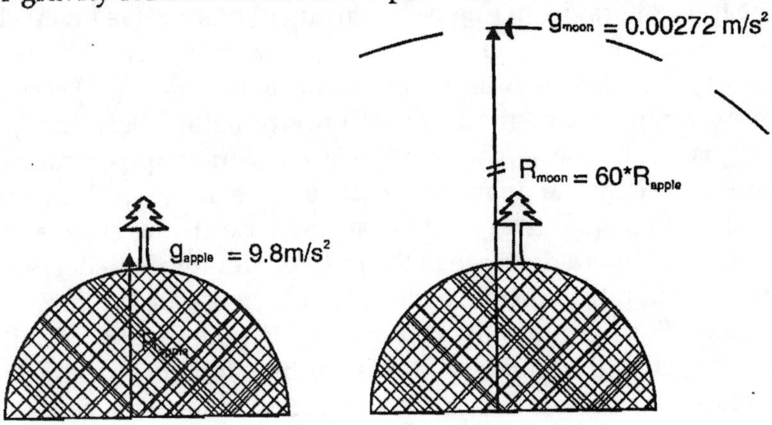

The relationship between the force of gravity (F_{grav}) between the earth and any other object and the distance which separates their centres (d) can be expressed by the following relationship

$$F_{grav} \sim \frac{1}{d^2}$$

where F_{grav} represents the force of gravity between two objects
~ means "proportional to"
d represents the distance separating the objects centers

Since the distance d is in the denominator of this relationship, it can be said that the force of gravity is inversely related to the distance. And since the distance is raised to the second power, it can be said that the force of gravity is inversely related to the square of the distance. This mathematical relationship is sometimes referred to as an inverse square law since one quantity depends inversely upon the square of the other quantity. The inverse square relation between the force of gravity and the distance of separation provided sufficient evidence for Newton's explanation of why gravity can be credited as the cause of both the falling apple's acceleration and the orbiting moon's acceleration.

Using Equations as a Guide to Thinking

The inverse square law proposed by Newton suggests that the force of gravity acting between any two objects is inversely proportional to the square of the separation distance between the object's centres. Altering the separation distance (d) results in an alteration in the force of gravity acting between the objects. Since the two quantities are inversely proportional, an increase in one quantity results in a decrease in the value of the other quantity. That is, an increase in the separation distance causes a decrease in the force of gravity and a decrease in the separation distance causes an increase in the force of gravity. Furthermore, the factor by which the force of gravity is changed is the square of the factor by which the separation distance is changed. So if the separation distance is doubled (increased by a factor of 2), then the force of gravity is decreased by a factor of four. And if the separation distance is tripled (increased by a factor of 3), then the force of gravity is decreased by a factor of nine (3 raised to the second power). Thinking of the force-distance relationship in this way involves using a mathematical relationship as a guide to thinking about how an alteration in one variable affects the other variable. Equations can be more than recipes for algebraic problem-solving; they can be *guides to thinking*. Check your understanding of the inverse square law as a guide to thinking by answering the following questions below. When finished, click the button to check your answers.

SATELLITE MOTION

A satellite is often thought of as being a projectile which is orbiting the Earth. But how can a projectile orbit the Earth? Doesn't a projectile accelerate towards the Earth under the influence of gravity? And as such, wouldn't any projectile ultimately fall towards the Earth and collide with the Earth, thus ceasing its orbit?

These are all good questions and represent stumbling blocks for many students of physics. We will discuss each question here. First, an orbiting satellite is a projectile in the sense that the only force acting upon an orbiting satellite is the force of gravity. Most Earth-orbiting satellites are orbiting at a distance high above the Earth such that their motion is unaffected by forces of air resistance. Indeed, a satellite is a projectile.

Second, a satellite is acted upon by the force of gravity and this force *does* accelerate it towards the Earth. In the absence of gravity a satellite would move in a straight line path tangent to the Earth. In the absence of any forces whatsoever, an object in motion (such as a satellite) would continue in motion with the same speed and in the same direction. This is the law of inertia. The force of gravity acts upon a high speed satellite to deviate its trajectory from a straight-line inertial path. Indeed, a satellite is accelerating towards the Earth due to the force of gravity.

Finally, a satellite does fall towards the Earth; only it never falls into the Earth. To understand this concept, we have to remind ourselves of the fact that the Earth is round; that is the Earth curves. In fact, scientists know that on average, the Earth curves approximately 5 meters downward for every 8000 meters along its horizon. If you were to look out horizontally along the horizon of the Earth for 8000 meters, you would observe that the Earth curves downwards below this straight-line path a distance of 5 meters. In order for a satellite to successfully orbit the Earth, it must travel a horizontal distance of 8000 meters before falling a vertical distance of 5 meters. A horizontally launched projectile falls a vertical distance of 5 meters in its first second of motion. To avoid hitting the Earth, an orbiting projectile must be launched with a horizontal speed of 8000 m/s. When launched at this speed, the projectile will fall towards the Earth with a trajectory which matches the curvature of the Earth. As such, the projectile will fall around the Earth, always accelerating towards it under the influence of gravity, yet never colliding into it since the Earth is constantly curving at the same rate. Such a projectile is an orbiting satellite.

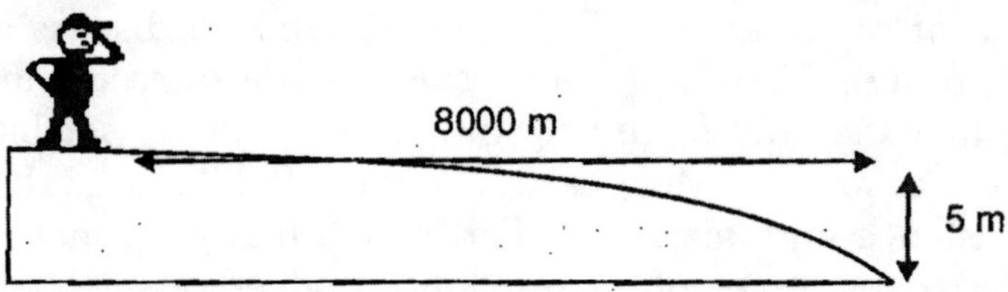

To further understanding the concept of a projectile orbiting around the Earth, consider the following thought experiment. Suppose that a very powerful cannon was mounted on top of a very tall mountain. Suppose that the mountain was so tall that any object set in motion from the mountaintop would be unaffected by air drag. Suppose that several cannonballs were fired from the cannon at various speeds - say speeds of 8000 m/s, less than 8000 m/s, and more than 8000 m/s. A cannonball launched with speeds less than 8000 m/s would eventually fall to the Earth. A cannonball launched with a speed of 8000 m/s would orbit the Earth in a circular path. Finally, a cannonball launched with a speed greater than 8000 m/s would orbit the Earth in an elliptical path. The animations below depict these ideas.

Important

Two final notes should be made about satellite motion. First, the 8000 m/s figure used in the above discussion applies to satellites launched from heights just above Earth's surface. Since gravitational influences decrease with the height above the Earth, the orbital speed required for a circular orbit is less than 8000 m/s at significantly greater heights above Earth's surface.

Second, there is an upper limit on the orbital speed of a satellite. If launched with two great of a speed, a projectile will escape Earth's gravitational influences and continue in motion without actually orbiting the Earth. Such a projectile will continue in motion until influenced by the gravitational influences of other celestial bodies.

NEWTON'S LAW OF UNIVERSAL GRAVITATION

Isaac Newton compared the acceleration of the moon to the acceleration of objects on earth. Believing that gravitational forces were responsible for each, Newton was able to draw an important conclusion about the dependence of gravity upon distance. This comparison led him to conclude that the force of gravitational attraction between the Earth and other objects is inversely proportional to the distance separating the earth's centre from the object's centre. But distance is not the only variable affecting the magnitude of a gravitational force. Consider Newton's famous equation

$$F_{net} = m \bullet a$$

Newton knew that the force which caused the apple's acceleration (gravity) must be dependent upon the mass of the apple. And since the force acting to cause the apple's downward acceleration also causes the earth's upward acceleration (Newton's third law), that force must also depend upon the mass of the earth. So for Newton, the force of gravity acting between the earth and any other object is directly proportional to the mass of the earth, directly proportional to the mass of the object, and inversely proportional to the square of the distance which separates the centres of the earth and the object.

But Newton's law of universal gravitation extends gravity beyond earth. Newton's law of universal gravitation is about the universality of gravity. Newton's place in the *Gravity Hall of Fame* is not due to his discovery of gravity, but rather due to his discovery that gravitation is universal. ALL objects attract each other with a force of gravitational attraction. Gravity is universal. This force of gravitational attraction is directly dependent upon the masses of both objects and inversely proportional to the square of the distance which separates their centres. Newton's conclusion about the magnitude of gravitational forces is summarized symbolically as

$$F_{grav} \sim \frac{m_1 * m_2}{d^2}$$

Where F_{grav} represents the force of gravity between two objects ~ means "proportional to" m_1 represents the mass of object 1, m_2 represents the mass of objects 2, d represents the distance separating the object's centres

Since the gravitational force is directly proportional to the mass of both interacting objects, more massive objects will attract each other with a greater gravitational force. So as the mass of either object increases, the force of gravitational attraction between them also increases. If the mass of one of the objects is doubled, then the force of gravity between them is doubled. If the mass of one of the objects is tripled, then the force of gravity between them is tripled. If the mass of both of the objects is doubled, then the force of gravity between them is quadrupled; and so on.

Since gravitational force is inversely proportional to the separation distance between the two interacting objects, more separation distance will result in weaker gravitational forces. So as two objects are separated from each other, the force of gravitational attraction between them also decreases. If the separation distance between two objects is doubled (increased by a factor of 2), then the force of gravitational attraction is decreased by a factor of 4. If the separation distance between any two objects is tripled, then the force of gravitational attraction is decreased by a factor of 9 (3 raised to the second power).

The proportionalities expressed by Newton's universal law of gravitation is represented graphically by the following illustration. Observe how the force of gravity is directly proportional to the product of the two masses and inversely proportional to the square of the distance of separation.

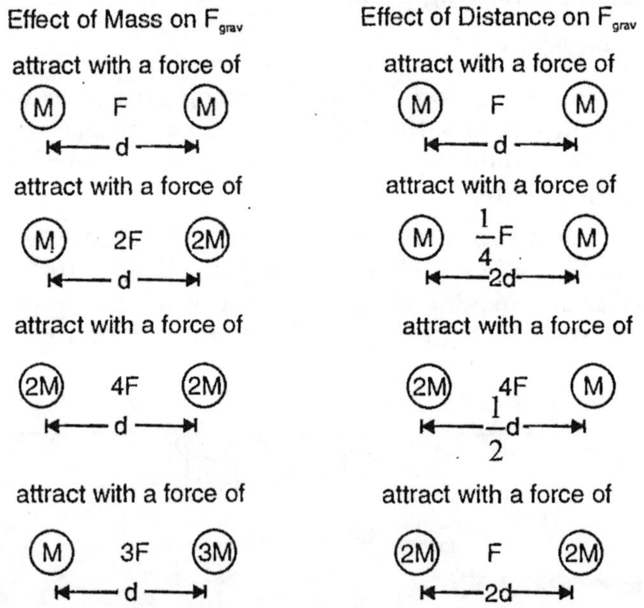

Another means of representing the proportionalities is to express the relationships in the form of an equation using a constant of proportionality. This equation is shown below.

$$F_{grav} = \frac{G * m_1 * m_2}{d^2}$$

where G represents the universal gravitation constant (G = 6.67×10^{-11} $N.m^2/kg^2$)

The constant of proportionality (G) in the above equation is known as the universal gravitation constant. The precise value of G was determined experimentally by Henry Cavendish in the century after Newton's death. The value of G is found to be

$$G = 6.673 \times 10^{-11} \text{ N m}^2/\text{kg}^2$$

The units on G may seem rather odd; nonetheless they are sensible. When the units on G are substituted into the equation above and multiplied by $m_1 \bullet m_2$ units and divided by d^2 units, the result will be Newtons - the unit of force.

Knowing the value of G allows us to calculate the force of gravitational attraction between any two objects of known mass and known separation distance. As a first example, consider the following problem.

Sample Problem #1

Determine the force of gravitational attraction between the earth (m = 5.98 × 10^{24} kg) and a 70-kg physics student if the student is standing at sea level, a distance of 6.38 x 10^6 m from earth's centre.

The solution of the problem involves substituting known values of G (6.673 × 10^{-11} N m^2/kg^2), m_1 (5.98 × 10^{24} kg), m_2 (70 kg) and d (6.38 x 10^6 m) into the universal gravitation equation and solving for F_{grav}. The solution is as follows:

$$F_{grav} = \frac{\left(6.673\times10^{-11}Nm^2/kg^2\right)\cdot\left(5.98\times10^{24}kg\right)\cdot(70kg)}{\left(6.38\times10^6 m\right)^2}$$

$$F_{grav} = 686N$$

Sample Problem #2

Determine the force of gravitational attraction between the earth (m = 5.98 x 10^{24} kg) and a 70-kg physics student if the student is in an airplane at 40000 feet above earth's surface. This would place the student a distance of 6.39 x 10^6 m from earth's centre.

The solution of the problem involves substituting known values of G (6.673 x 10^{-11} N m^2/kg^2), m_1 (5.98 x 10^{24} kg), m_2 (70 kg) and d (6.39 x 10^6 m) into the universal gravitation equation and solving for F_{grav}. The solution is as follows:

$$Fgrav = \frac{\left(6.673\times10^{-11}Nm^2/kg^2\right)\cdot\left(5.98\times10^{24}kg\right)\cdot(70kg)}{\left(6.39\times10^6 m\right)^2}$$

$F_{grav} = 684N$

Two general conceptual comments can be made about the results of the two sample calculations above. First, observe that the force of gravity acting upon the student (a.k.a. the student's weight) is less on an airplane at 40 000 feet than at sea level. This illustrates the inverse relationship between separation distance and the force of gravity (or in this case, the weight of the student). The student weighs less at the higher altitude. However, a mere change of 40 000 feet further from the centre of the Earth is virtually negligible. This altitude change altered the student's weight changed by 2 N which is much less than 1% of the original weight. A distance of 40 000 feet (from the earth's surface to a high altitude airplane) is not very far when compared to a distance of 6.38×10^6 m (equivalent to nearly 20 000 000 feet from the centre of the earth to the surface of the earth). This alteration of distance is like *a drop in a bucket* when compared to the large radius of the Earth. As shown in the diagram below, distance of separation becomes much more influential when a significant variation is made.

Force of Gravitational Attraction towards Earth
for a 70-kg Physics Student at Various Locations

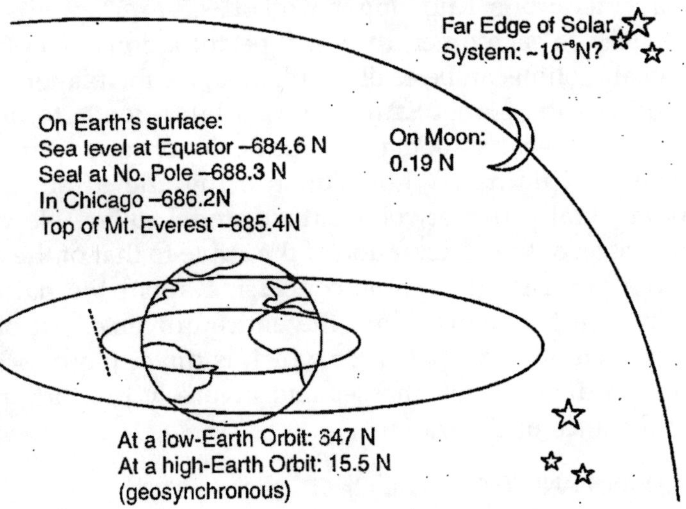

The second conceptual comment to be made about the above sample calculations is that the use of Newton's universal gravitation equation to calculate the force of gravity (or weight) yields the same result.

$$F_{grav} = m \bullet g = (70\ kg) \bullet (9.8\ m/s^2) = 686\ N$$

Both equations accomplish the same result because the value of g is equivalent to the ratio of $(G \bullet M_{earth})/(R_{earth})^2$.

THE UNIVERSALITY OF GRAVITY

Gravitational interactions do not simply exist between the earth and other objects; and not simply between the sun and other planets. Gravitational interactions exist between all objects with an intensity which is directly proportional to the product of their masses. So as you sit in your seat in the physics classroom, you are gravitationally attracted to your lab partner, to the desk you are working at, and even to your physics book. Newton's revolutionary idea was that gravity is universal - ALL objects attract in proportion to the product of their masses. Gravity is universal. Of course, most gravitational forces are so minimal to be noticed. Gravitational forces are only recognizable as the masses of objects become large.

Today, Newton's law of universal gravitation is a widely accepted theory. It guides the efforts of scientists in their study of planetary orbits. Knowing that all objects exert gravitational influences on each other, the small perturbations in a planet's elliptical motion can be easily explained. As the planet Jupiter approaches the planet Saturn in its orbit, it tends to deviate from its otherwise smooth path; this deviation, or perturbation, is easily explained when considering the affect of the gravitational pull between Saturn and Jupiter. Newton's comparison of the acceleration of the apple to that of the moon led to a surprisingly simple conclusion about the nature of gravity which is woven into the entire universe. All objects attract each other with a force which is directly proportional to the product of their masses and inversely proportional to their distance of separation

CAVENDISH AND THE VALUE OF G

Isaac Newton's law of universal gravitation proposed that the gravitational attraction between any two objects is directly proportional to the product of their masses and inversely proportional to the distance between their centres. In equation form, this is often expressed as follows:

$$F_{grav} = \frac{G * m_1 * m_2}{d^2}$$

The constant of proportionality in this equation is G - the universal gravitation constant. The value of G was not experimentally determined until nearly a century later (1798) by Lord Henry Cavendish using a torsion balance.

Cavendish's apparatus for experimentally determining the value of G involved a light, rigid rod about 2-feet long. Two small lead spheres were attached to the ends of the rod and the rod was suspended by a thin wire. When the rod becomes twisted, the torsion of the wire begins to exert a torsional force which is proportional to the angle of rotation of the rod. The more twist of the wire; the more the system pushes *backwards* to restore itself towards the original position. Cavendish had calibrated his instrument to determine the relationship between the angle of rotation and the amount of torsional force. A diagram of the apparatus is shown below.

Cavendish's Torsion Balance

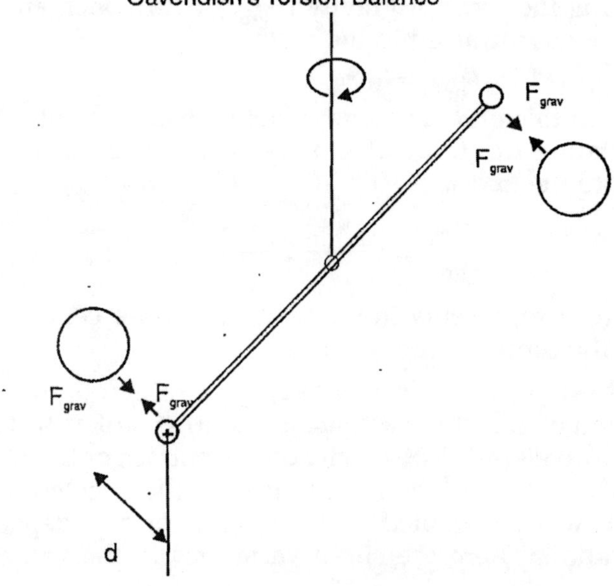

Cavendish then brought two large lead spheres near the smaller spheres attached to the rod. Since all masses attract, the large spheres exerted a gravitational force upon the smaller spheres and twisted the rod a measurable amount. Once the torsional force balanced the gravitational force, the rod and spheres came to rest and Cavendish was able to determine the gravitational force of attraction between the masses. By measuring m_1, m_2, d and F_{grav}, the value of G could be determined. Cavendish's measurements resulted in an experimentally determined value of 6.75×10^{-11} N m^2/kg^2. Today, the currently accepted value is 6.67259×10^{-11} N m^2/kg^2.

The value of G is an extremely small numerical value. Its smallness accounts for the fact that the force of gravitational attraction is only appreciable for objects with large mass. While two students will indeed exert gravitational forces upon each other, these forces are too small to be noticeable. Yet if one of the students is replaced with a planet, then the gravitational force between the other student and the planet becomes noticeable.

THE VALUE OF G

The Physics Classroom, an equation was given for determining the force of gravity (F_{grav}) with which an object of mass m was attracted to the earth

$$F_{grav} = m^*g$$

Now in this unit, a second equation has been introduced for calculating the force of gravity with which an object is attracted to the earth.

$$F_{grav} = \frac{G * m_{earth} * m}{d^2}$$

where d represents the distance from the centre of the object to the centre of the earth.

In the first equation above, g is referred to as the acceleration of gravity. It's value is 9.8 m/s^2 on Earth. That is to say, the acceleration of gravity on the surface of the earth at sea level is 9.8 m/s^2. When discussing the acceleration of gravity, it was mentioned that the value of g is dependent upon location. There are slight variations in the value of g about earth's surface. These variations result from the varying density of the geologic structures below each specific surface location. They also result from the fact that the earth is not truly spherical; the earth's surface is further from its centre at the equator than it is at the poles. This would result in larger g values at the poles. As one proceeds further from earth's surface - say into a location of orbit about the earth - the value of g changes still.

To understand why the value of g is so location dependent, we will use the two equations above to derive an equation for the value of g. First, both expressions for the force of gravity are set equal to each other.

$$m^*g = \frac{G * M_{earth} * m}{d^2}$$

Now observe that the mass of the object - m - is present on both sides of the equal sign. Thus, m can be canceled from the equation. This leaves us with an equation for the acceleration of gravity.

$$g = \frac{G * M_{earth}}{d^2}$$

The above equation demonstrates that the acceleration of gravity is dependent upon the mass of the earth (approx. 5.98×10^{24} kg) and the distance (d) that an object is from the centre of the earth. If the value 6.38×10^6 m (a typical earth radius value) is used for the distance from Earth's centre, then g will be calculated to be 9.8 m/s^2. And of course, the value of g will change as an object is moved further from Earth's centre. For instance, if an object were moved to a location which is two earth-radii from the centre of the earth - that is, two times 6.38×10^6 m - then a significantly different value of g will be found. As shown below, at twice the distance from the centre of the earth, the value of g becomes 2.45 m/s^2.

$$g = \frac{\left(6.673 \times 10^{-11} N.m^2 / kg^2\right) * \left(5.98 \times 10^{24} kg\right)}{\left(1.276 \times 10^7 m\right)^2}$$

The table below shows the value of g at various locations from Earth's centre.

Location	Distance from Earth's centre (m)	Value of gm/s^2
Earth's surface	6.38×10^6 m	9.8
1000 km above surface	7.38×10^6 m	7.33
2000 km above surface	8.38×10^6 m	5.68
3000 km above surface	9.38×10^6 m	4.53
4000 km above surface	1.04×10^7 m	3.70
5000 km above surface	1.14×10^7 m	3.08
6000 km above surface	1.24×10^7 m	2.60
7000 km above surface	1.34×10^7 m	2.23
8000 km above surface	1.44×10^7 m	1.93
9000 km above surface	1.54×10^7 m	1.69
10000 km above surface	1.64×10^7 m	1.49
50000 km above surface	5.64×10^7 m	0.13

As is evident from both the equation and the table above, the value of g varies inversely with the distance from the centre of the earth. In fact, the variation in g with distance follows an inverse square law where g is inversely proportional to the distance from earth's centre. This inverse square relationship means that as the distance is doubled, the value of g decreases by a factor of 4. As the distance is tripled, the value of g decreases by a factor of 9; and so on. This inverse square relationship is depicted in the graphic at the right.

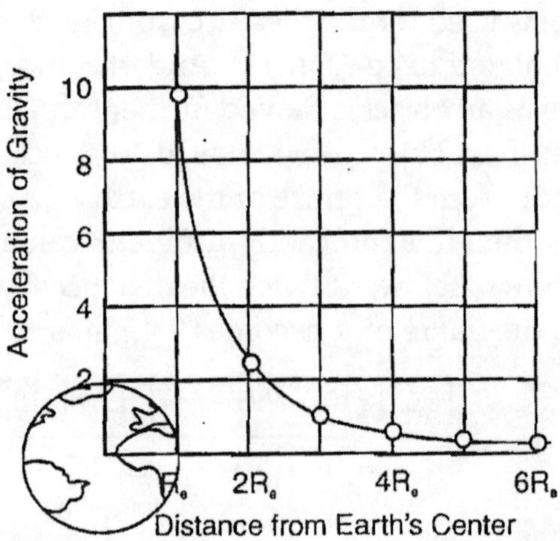

The same equation used to determine the value of g on Earth' surface can also be used to determine the acceleration of gravity on the surface of other planets. The value of g on any other planet can be calculated from the mass of the planet and the radius of the planet. The equation takes the following form:

$$g = \frac{G * M_{planet}}{R_{planet}^{2}}$$

Using this equation, the following acceleration of gravity values can be calculated for the various planets.

Planet	Radius (m)	Mass (kg)	g (m/s^2)
Mercury	2.43×10^6	3.2×10^{23}	3.61
Venus	6.073×10^6	4.88×10^{24}	8.83
Mars	3.38×10^6	6.42×10^{23}	3.75
Jupiter	6.98×10^7	1.901×10^{27}	26.0
Saturn	5.82×10^7	5.68×10^{26}	11.2
Uranus	2.35×10^7	8.68×10^{25}	10.5
Neptune	2.27×10^7	1.03×10^{26}	13.3
Pluto	1.15×10^6	1.2×10^{22}	0.61

The acceleration of gravity of an object is a measurable quantity. Yet emerging from Newton's universal law of gravitation is a prediction which states that its value is dependent upon the mass of the Earth and the distance the object is from the Earth's centre. The value of g is independent of the mass of the object and only dependent upon *location* - the planet the object is on and the distance from the centre of that planet.

PLANETARY AND SATELLITE MOTION

Kepler's Three Laws

In the early 1600s, Johannes Kepler proposed three laws of planetary motion. Kepler was able to summarize the carefully collected data of his mentor - Tycho Brahe - with three statements which described the motion of planets in a sun-centreed solar system. Kepler's efforts to explain the underlying reasons for such motions are no longer accepted;

nonetheless, the actual laws themselves are still considered an accurate description of the motion of any planet and any satellite.

Kepler's three laws of planetary motion can be described as follows:

- The path of the planets about the sun are elliptical in shape, with the centre of the sun being located at one focus. (The Law of Ellipses)
- An imaginary line drawn from the centre of the sun to the centre of the planet will sweep out equal areas in equal intervals of time. (The Law of Equal Areas)
- The ratio of the squares of the periods of any two planets is equal to the ratio of the cubes of their average distances from the sun. (The Law of Harmonies)

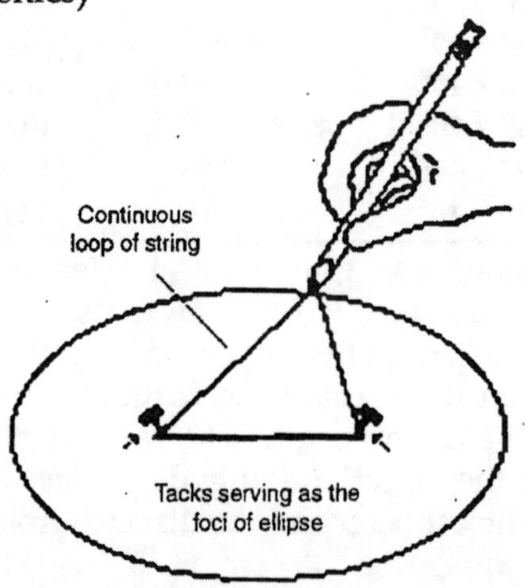

Kepler's first law - sometimes referred to as the law of ellipses - explains that planets are orbiting the sun in a path described as an ellipse. An ellipse can easily be constructed using a pencil, two tacks, a string, a sheet of paper and a piece of cardboard. Tack the sheet of paper to the cardboard using the two tacks. Then tie the string into a loop and wrap the loop around the two tacks. Take your pencil and pull the string until the pencil and two tacks make a triangle (see diagram at the right). Then begin to trace out a path with the pencil, keeping the string wrapped tightly around the tacks. The resulting shape will be an ellipse. An ellipse is a special curve in which the sum of the distances from every point on the curve to two other points is a constant. The two other points are known as the foci of the ellipse. The closer together which these points are, the more closely that the ellipse resembles the shape of a circle. In fact, a circle is the special case of an ellipse in which the two foci are at the same location. Kepler's first law is rather simple - all planets orbit the sun in a path which resembles an ellipse, with the sun being located at one of the foci of that ellipse.

The Law of Equal Areas

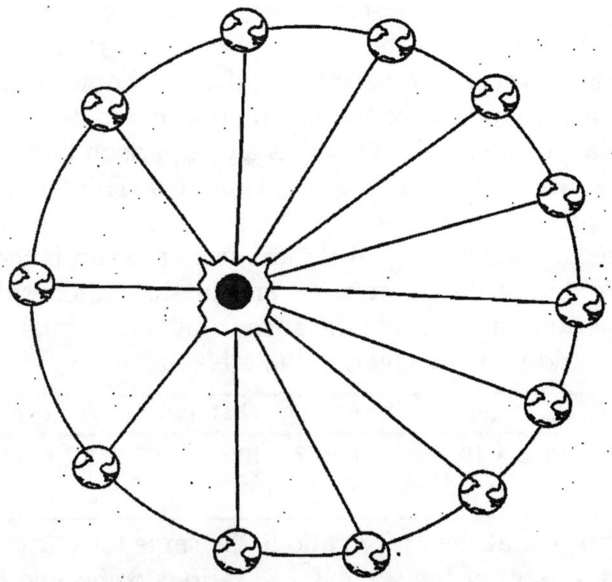

An imaginary line drawn from the sun to any planet sweeps out equal areas in equal amounts of time.

Kepler's second law - sometimes referred to as the law of equal areas - describes the speed at which any given planet will move while orbiting the sun. The speed at which any planet moves through space is constantly changing. A planet moves fastest when it is closest to the sun and slowest when it is furthest from the sun. Yet, if an imaginary line were drawn from the centre of the planet to the centre of the sun, that line would sweep out the same area in equal periods of time. For instance, if an imaginary line were drawn from the earth to the sun, then the area swept out by the line in every 31-day month would be the same. This is depicted in the diagram below. As can be observed in the diagram, the areas formed when the earth is closest to the sun can be approximated as a wide but short triangle; whereas the areas formed when the earth is farthest from the sun can be approximated as a narrow but long triangle. These areas are the sam e size. Since the *base* of these triangles are longer when the earth is furthest from the sun, the earth would have to be moving more slowly in order for this imaginary area to be the same size as when the earth is closest to the sun.

Kepler's third law - sometimes referred to as the law of harmonies - compares the orbital period and radius of orbit of a planet to those of other planets. Unlike Kepler's first and second laws which describe the motion characteristics of a single planet, the third law makes a comparison between the motion characteristics of different planets. The comparison being made is that the ratio of the squares of the periods to the cubes of their average distances from the sun is the same for every one of the planets. As an illustration, consider the orbital period and average distance from sun (orbital radius) for Earth and mars as given in the table below.

Planet	Period(s)	Average Dist. (m)	$T^2/R^3(s^2/m^3)$
Earth	3.156×10^7 s	1.4957×10^{11}	2.977×10^{-19}
Mars	5.93×10^7 s	2.278×10^{11}	2.975×10^{-19}

Observe that the T^2/R^3 ratio is the same for Earth as it is for mars. In fact, if the same T^2/R^3 ratio is computed for the other planets, it can be found that this ratio is nearly the same value for all the planets (see table below). Amazingly, every planet has the same T^2/R^3 ratio.

Planet	Period(yr)	Ave.Dist. (au)	$T^2/R^3(yr^2/au^3)$
Mercury	0.241	0.39	0.98
Venus	.615	0.72	1.01
Earth	1.00	1.00	1.00
Mars	1.88	1.52	1.01
Jupiter	11.8	5.20	0.99

Saturn	29.5	9.54	1.00
Uranus	84.0	19.18	1.00
Neptune	165	30.06	1.00
Pluto	248	39.44	1.00

(*Note:* The average distance value is given in astronomical units where 1 a.u. is equal to the distance from the earth to the sun - 1.4957×10^{11} m. The orbital period is given in units of earth-years where 1 earth year is the time required for the earth to orbit the sun - 3.156×10^{7} seconds.)

Kepler's third law provides an accurate description of the period and distance for a planet's orbits about the sun. Additionally, the same law which describes the T^2/R^3 ratio for the planets' orbits about the sun also accurately describes the T^2/R^3 ratio for any satellite (whether a moon or a man-made satellite) about any planet. There is something much deeper to be found in this T^2/R^3 ratio - something which must relate to basic fundamental principles of motion.

How did Newton Extend His Notion of Gravity to Explain Planetary Motion?

Newton's comparison of the acceleration of the moon to the acceleration of objects on earth allowed him to establish that the moon is held in a circular orbit by the force of gravity - a force which is inversely dependent upon the distance between the two objects' centres. Establishing gravity as the cause of the moon's orbit does not necessarily establish that gravity is the cause of the planet's orbits. How then did Newton provide credible evidence that the force of gravity is meets the centripetal force requirement for the elliptical motion of planets?

Johannes Kepler proposed three laws of planetary motion. His Law of Harmonies suggested that the ratio of the period of orbit squared (T^2) to the mean radius of orbit cubed (R^3) is the same value k for all the planets which orbit the sun. Known data for the orbiting planets suggested the following average ratio:

$$k = 2.97 \times 10^{-19}\ s^2/m^3 = (T^2)/(R^3)$$

Newton was able to combine the law of universal gravitation with circular motion principles to show that if the force of gravity provides the centripetal force for the planets' nearly circular orbits, then a value of $2.97 \times 10^{-19}\ s^2/m^3$ could be predicted for the T^2/R^3 ratio. Here is the reasoning employed by Newton:

Consider a planet with mass M_{planet} to orbit in nearly circular motion about the sun of mass M_{Sun}. The net centripetal force acting upon this orbiting planet is given by the relationship

$$F_{net} = (M_{planet} * v^2) / R$$

This net centripetal force is the result of the gravitational force which attracts the planet towards the sun, and can be represented as

$$F_{grav} = (G^* M_{planet} * M_{Sun}) / R^2$$

Since $F_{grav} = F_{net}$, the above expressions for centripetal force and gravitational force are equal. Thus,

$$(M_{planet} * v^2) / R = (G^* M_{planet} * M_{Sun}) / R^2$$

Since the velocity of an object in nearly circular orbit can be approximated as $v = (2*pi*R) / T$,

$$v^2 = (4 * pi^2 * R^2) / T^2$$

Substitution of the expression for v^2 into the equation above yields,

$$(M_{planet} * 4 * pi^2 * R^2) / (R \bullet T^2) = (G^* M_{planet} * M_{Sun}) / R^2$$

By cross-multiplication and simplification, the equation can be transformed into

$$T^2 / R^3 = (M_{planet} * 4 * pi^2) / (G^* M_{planet} * M_{Sun})$$

The mass of the planet can then be canceled from the numerator and the denominator of the equation's right-side, yielding

$$T^2 / R^3 = (4 * pi^2) / (G * M_{Sun})$$

The right side of the above equation will be the same value for every planet regardless of the planet's mass. Subsequently, it is reasonable that the T^2/R^3 ratio would be the same value for all planets if the force which holds the planets in their orbits is the force of gravity. Newton's universal law of gravitation predicts results which were consistent with known planetary data and provided a theoretical explanation for Kepler's Law of Harmonies.

CIRCULAR MOTION PRINCIPLES FOR SATELLITES

A satellite is any object which is orbiting the earth, sun or other massive body. Satellites can be categorized as natural satellites or man-made satellites. The moon, the planets and comets are examples of natural satellites. Accompanying the orbit of natural satellites are a host of satellites launched from earth for purposes of communication, scientific research, weather forecasting, intelligence, etc. Whether a moon, a planet, or some man-made satellite, every satellite's motion is governed by the same physics principles and described by the same mathematical equations.

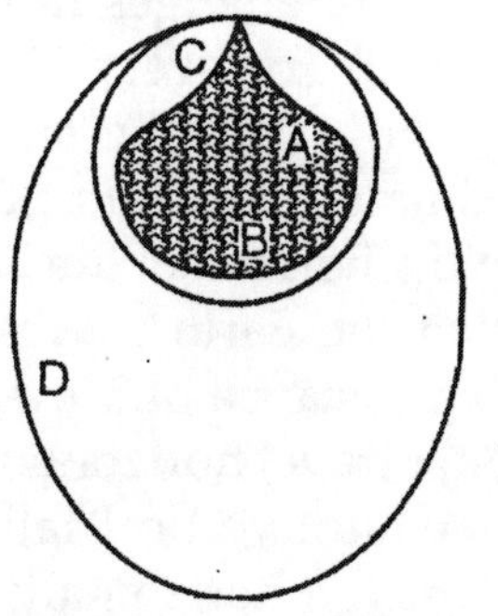

A satellite is a projectile which falls around the earth instead of into it.

The fundamental principle to be understood concerning satellites is that a satellite is a projectile. That is to say, a satellite is an object upon which the only force is gravity. Once launched into orbit, the only force governing the motion of a satellite is the force of gravity. Newton was the first to theorize that a projectile launched with sufficient speed would actually orbit the earth. Consider a projectile launched horizontally from the top of the legendary *Newton's Mountain* - at a location high above the influence of air drag. As the projectile moves horizontally in a direction tangent to the earth, the force of gravity would pull it downward. If the launch speed was too small, it would eventually fall to earth. The diagram at the right resembles that found in Newton's original writings. Paths A and B illustrate the path of a projectile with insufficient launch speed for orbital motion.

But if launched with sufficient speed, the projectile would fall towards the earth at the same rate that the earth curves. This would cause the projectile to stay the same height above the earth and to orbit in a circular path (such as path C). And at even greater launch speeds, a cannonball would once more orbit the earth, but now in an elliptical path (as in path D). At every point along its trajectory, a satellite is falling toward the earth. Yet because the earth curves, it never reaches the earth.

So what launch speed does a satellite need in order to orbit the earth? The answer emerges from a basic fact about the curvature of the earth. For every 8000 meters measured along the horizon of the earth, the earth's surface curves downward by approximately 5 meters. So if you were to look out horizontally along the horizon of the Earth for 8000 meters, you would observe that the Earth curves downwards below this straight-line path a distance of 5 meters. For a projectile to orbit the earth, it must travel horizontally a distance of 8000 meters for every 5 meters of vertical fall. It so happens that the vertical distance which a horizontally launched projectile would fall in its first second is approximately 5 meters ($0.5*g*t^2$). For this reason, a projectile launched horizontally with a speed of about 8000 m/s will be capable of orbiting the earth in a circular path. This assumes that it is launched above the surface of the earth and encounters negligible atmospheric drag. As the projectile travels tangentially a distance of 8000 meters in 1 second, it will drop approximately 5 meters towards the earth. Yet, the projectile will remain the same distance above the earth due to the fact that the earth curves at the same rate that the projectile falls. If shot with a speed greater than 8000 m/s, it would orbit the earth in an elliptical path.

Velocity, Acceleration and Force Vectors

The motion of an orbiting satellite can be described by the same motion characteristics as any object in circular motion. The velocity of the satellite would be directed tangent to the circle at every point along its path. The acceleration of the satellite would be directed towards the centre of the circle - towards the central body which it is orbiting. And this acceleration is caused by a net force which is directed inwards in the same direction as the acceleration.

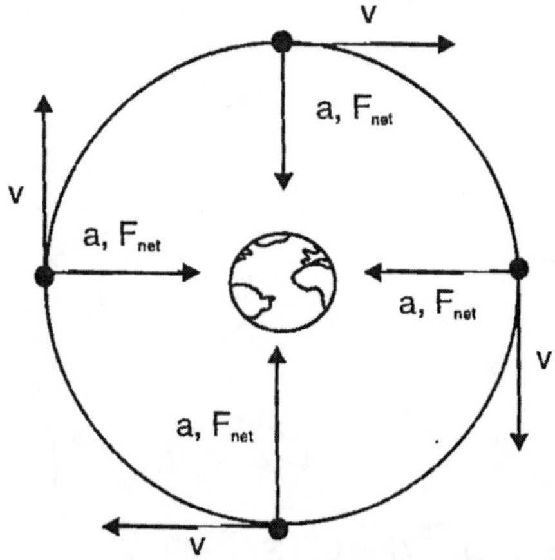

Satellites encounter inward forces and accelerations and tangential velocities.

This centripetal force is supplied by gravity - the force which universally acts at a distance between any two objects which have mass. Were it not for this force, the satellite in motion would continue in motion at the same speed and in the same direction. It would follow its inertial, straight-line path. Like any projectile, gravity alone influences the satellite's trajectory such that it always falls below its straight-line, inertial path. This is depicted in the diagram below. Observe that the inward net force pushes (or pulls) the satellite (denoted by blue circle) inwards relative to its straight-line path tangent to the circle. As a result, after the first interval of time, the satellite is positioned at position 1 rather than position 1'. In the next interval of time, the same satellite would travel tangent to the circle in the absence of gravity and be at position 2'; but because of the inward force the satellite has moved to position 2 instead. In the next interval of time, the same satellite has moved inward to position 3 instead of tangentially to position 3'. This same reasoning can be repeated to explain how the inward force causes the satellite to fall towards the earth without actually falling into it.

An Orbiting Satellite Requires a Centripetal Force

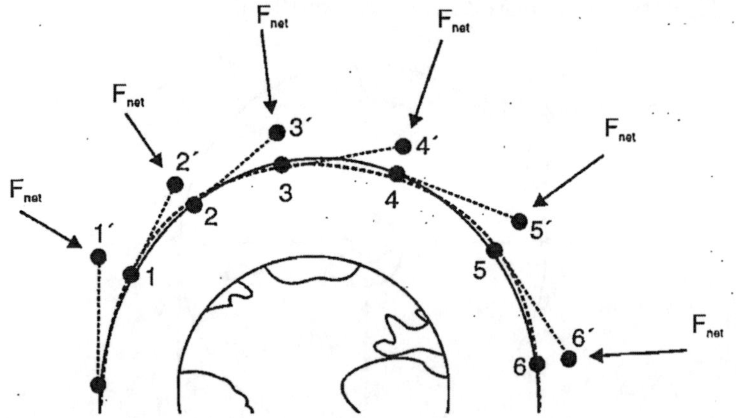

ELLIPTICAL ORBITS OF SATELLITES

Occasionally satellites will orbit in paths which can be described as ellipses. In such cases, the central body is located at one of the foci of the ellipse. Similar motion characteristics apply for satellites moving in elliptical paths. The velocity of the satellite is directed tangent to the ellipse. The acceleration of the satellite is directed towards the focus of the ellipse. And in accord with Newton's second law of motion, the net force acting upon the satellite is directed in the same direction as the acceleration - towards the focus of the ellipse.

Once more, this net force is supplied by the force of gravitational attraction between the central body and the orbiting satellite. In the case of elliptical paths, there is a component of force in the same direction as (or opposite direction as) the motion of the object. Such a component of force can cause the satellite to either speed up or slow down in addition to changing directions. So unlike uniform circular motion, the elliptical motion of satellites is not characterized by a constant speed.

In summary, satellites are projectiles which orbit around a central massive body instead of falling into it. Being projectiles, they are acted upon by the force of gravity - a universal force which acts over even large distances between any two masses. The motion of satellites, like any projectile, are governed by Newton's laws of motion. For this reason, the mathematics of these satellites emerges from an application of Newton's universal law of gravitation to the mathematics of circular motion.

Mathematics of Satellite Motion

The motion of objects are governed by Newton's laws. The same simple laws which govern the motion of objects on earth also extend to the *heavens* to govern the motion of planets, moons, and other satellites. The mathematics which describes a satellite's motion are the same mathematics presented for circular motion. We will be concerned with the variety of mathematical equations which describe the motion of satellites.

Consider a satellite with mass M_{sat} orbiting a central body with a mass of mass $M_{Central}$. The central body could be a planet, the sun or some other large mass capable of causing sufficient acceleration on a less massive nearby object. If the satellite moves in circular motion, then the net centripetal force acting upon this orbiting satellite is given by the relationship

$$F_{net} = (M_{sat} \bullet v^2) / R$$

This net centripetal force is the result of the gravitational force which attracts the satellite towards the central body and can be represented as

$$F_{grav} = (G \bullet M_{sat} \bullet M_{Central}) / R^2$$

Since $F_{grav} = F_{net}$, the above expressions for centripetal force and gravitational force can be set equal to each other. Thus,

$$(M_{sat} \bullet v^2) / R = (G \bullet M_{sat} \bullet M_{Central}) / R^2$$

Observe that the mass of the satellite is present on both sides of the equation; thus it can be canceled by dividing through by M_{sat}. Then both sides of the equation can be multiplied by R, leaving the following equation.

$$v^2 = (G \bullet M_{Central}) / R$$

Taking the square root of each side, leaves the following equation for the velocity of a satellite moving about a central body in circular motion

$$v = \sqrt{\frac{G^* M_{central}}{R}}$$

where G is 6.673×10^{-11} $N \bullet m^2/kg^2$, $M_{central}$ is the mass of the central body about which the satellite orbits, and R is the radius of orbit for the satellite.

Similar reasoning can be used to determine an equation for the acceleration of our satellite that is expressed in terms of masses and radius of orbit. The acceleration value of a satellite is equal to the acceleration of gravity of the satellite at whatever location which it is orbiting.The equation for the acceleration of gravity was given as

$$g = (G \bullet M_{central})/R^2$$

Thus, the acceleration of a satellite in circular motion about some central body is given by the following equation

$$a = \frac{G^* M_{central}}{R^2}$$

where G is 6.673×10^{-11} $N \bullet m^2/kg^2$, $M_{central}$ is the mass of the central body about which the satellite orbits, and R is the average radius of orbit for the satellite.

The final equation which is useful in describing the motion of satellites is Newton's form of Kepler's third law. Since the logic behind the development of the equation has been presented elsewhere, only the equation will be presented here. The period of a satellite (T) and the mean distance from the central body (R) are related by the following equation:

$$\frac{T^2}{R^3} = \frac{4 * \pi^2}{G^* M_{central}}$$

where T is the period of the satellite, R is the average radius of orbit for the satellite (distance from centre of central planet), and G is $6.673 \times 10^{-11}\ N \bullet m^2/kg^2$.

There is an important concept evident in all three of these equations - the period, speed and the acceleration of an orbiting satellite are not dependent upon the mass of the satellite.

$$v = \sqrt{\frac{G * M_{central}}{R}}, \quad a = \frac{G * M_{central}}{R^2}, \quad \frac{T^2}{R^3} = \frac{4 * \pi^2}{G * M_{central}}$$

None of these three equations has the variable $M_{satellite}$ in them. The period, speed and acceleration of a satellite is only dependent upon the radius of orbit and the mass of the central body which the satellite is orbiting. Just as in the case of the motion of projectiles on earth, the mass of the projectile has no affect upon the acceleration towards the earth and the speed at any instant. When air resistance is negligible and only gravity is present, the mass of the moving object becomes a non-factor. Such is the case of orbiting satellites.

Example Problems

To illustrate the usefulness of the above equations, consider the following practice problems.

Practice Problem #1

A satellite wishes to orbit the earth at a height of 100 km (approximately 60 miles) above the surface of the earth. Determine the speed, acceleration and orbital period of the satellite. (Given: $M_{earth} = 5.98 \times 10^{24}$ kg, $R_{earth} = 6.37 \times 10^6$ m)

Like most problems in physics, this problem begins by identifying known and unknown information and selecting the appropriate equation capable of solving for the unknown. For this problem, the knowns and unknowns are listed below.

Given/Known:

$R = R_{earth} + \text{height} = 6.47 \times 10^6$ m

$M_{earth} = 5.98 \text{x} 10^{24}$ kg

$G = 6.673 \times 10^{-11}\ N\ m^2/kg^2$Unknown:

v = ???

a = ???

T = ???

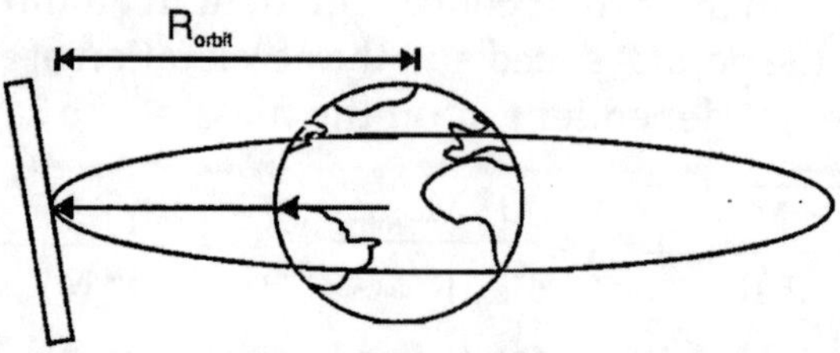

Note that the radius of a satellite's orbit can be found from the knowledge of the earth's radius and the height of the satellite above the earth. As shown in the diagram at the right, the radius of orbit for a satellite is equal to the sum of the earth's radius and the height above the earth. These two quantities can be added to yield the orbital radius. In this problem, the 100 km must first be converted to 100 000 m before being added to the radius of the earth. The equations needed to determine the unknown are listed above. We will begin by determining the orbital speed of the satellite using the following equation:

$v = SQRT\ [\ (G \bullet M_{Central})\ /\ R]$

The substitution and solution are as follows:

$v = SQRT\ [\ (6.673 \times 10^{-11}\ N\ m^2/kg^2) \bullet (5.98 \times 10^{24}\ kg)\ /\ (6.47 \times 10^6\ m)\ ]$

$v = 7.85 \times 10^3\ m/s$

The acceleration can be found from either one of the following equations:

$a = (G \bullet M_{central})/R^2 (2)\ a = v^2/R$

Equation is a general equation for circular motion. Either equation can be used to calculate the acceleration. The use of equation will be demonstrated here.

$a = (G \bullet M_{central})/R^2$

$a = (6.673 \times 10^{-11}\ N\ m^2/kg^2) \bullet (5.98 \times 10^{24}\ kg)\ /\ (6.47 \times 10^6\ m)^2$

$a = 9.53\ m/s^2$

Observe that this acceleration is slightly less than the 9.8 m/s^2 value expected on earth's surface. The increased distance from the centre of the earth lowers the value of g.

Finally, the period can be calculated using the following equation:

$$\frac{T^2}{R^3} = \frac{4 * \pi^2}{G * M_{central}}$$

The equation can be rearranged to the following form

$T = SQRT\ [(4 \bullet pi^2 \bullet R^3) / (G^*M_{central})]$

The substitution and solution are as follows:

$T = SQRT\ [(4 \bullet (3.1415)^2 \bullet (6.47 \times 10^6\ m)^3) / (6.673 \times 10^{-11}\ N\ m^2/kg^2) \bullet (5.98 \times 10^{24}\ kg)\ ]$

$T = 5176\ s = 1.44\ hrs$

Practice Problem #2

The period of the moon is approximately 27.2 days (2.35×10^6 s). Determine the radius of the moon's orbit and the orbital speed of the moon. (Given: $M_{earth} = 5.98 \times 10^{24}$ kg, $R_{earth} = 6.37 \times 10^6$ m)

Like Practice Problem #2, this problem begins by identifying known and unknown values. These are shown below.

Given/Known:

$T = 2.35 \times 10^6$ s

$M_{earth} = 5.98 \times 10^{24}$ kg

$G = 6.673 \times 10^{-11}\ N\ m^2/kg^2$Unknown:

R = ???

v = ???

The radius of orbit can be calculated using the following equation:

$$\frac{T^2}{R^3} = \frac{4 * \pi^2}{G * M_{central}}$$

The equation can be rearranged to the following form

$R^3 = [(T^2 \bullet G \bullet M_{central}) / (4 \bullet pi^2)]$

The substitution and solution are as follows:

$R^3 = [((2.35x10^6 \text{ s})^2 \bullet (6.673 \times 10^{-11} \text{ N m}^2/\text{kg}^2) \bullet (5.98x10^{24} \text{ kg})) / (4 \bullet (3.1415)^2)]$

$R^3 = 5.58 \times 10^{25} \text{ m}^3$

By taking the cube root of 5.58 x 10^{25} m^3, the radius can be determined as follows:

$R = 3.82 \times 10^8 \text{ m}$

The orbital speed of the satellite can be computed from either of the following equations:

$v = SQRT [(G \bullet M_{Central}) / R](2)$ $v = (2 \bullet pi \bullet R)/T$

Equation was derived above. Equation is a general equation for circular motion. Either equation can be used to calculate the orbital speed; the use of equation will be demonstrated here. The substitution of values into this equation and solution are as follows:

$v = SQRT [(6.673 \times 10^{-11} \text{ N m}^2/\text{kg}^2)*(5.98x10^{24} \text{ kg}) / (3.82 \times 10^8 \text{ m})]$

$v = 1.02 \times 10^3 \text{ m/s}$

Practice Problem #3

A geosynchronous satellite is a satellite which orbits the earth with an orbital period of 24 hours, thus matching the period of the earth's rotational motion. A special class of geosynchronous satellites is a geostationary satellite. A geostationary satellite orbits the earth in 24 hours along an orbital path which is parallel to an imaginary plane drawn through the Earth's equator. Such a satellite appears permanently fixed above the same location on the Earth. If a geostationary satellite wishes to orbit the earth in 24 hours (86400 s), then how high above the earth's surface must it be located. (Given: $M_{earth} = 5.98x10^{24}$ kg, $R_{earth} = 6.37 \times 10^6$ m)

Just as in the previous problem, the solution begins by the identification of the known and unknown values. This is shown below.

Given/Known:

T = 86400 s

M_{earth} = 5.98x10^{24} kg

R_{earth} = 6.37 x 10^6 m

G = 6.673 x 10^{-11} N m^2/kg^2Unknown:

h = ???

The unknown in this problem is the height (h) of the satellite above the surface of the earth. Yet there is no equation with the variable h. The solution then involves first finding the radius of orbit and using this R value and the R of the earth to find the height of the satellite above the earth. As shown in the diagram at the right, the radius of orbit for a satellite is equal to the sum of the earth's radius and the height above the earth. The radius of orbit can be found using the following equation:

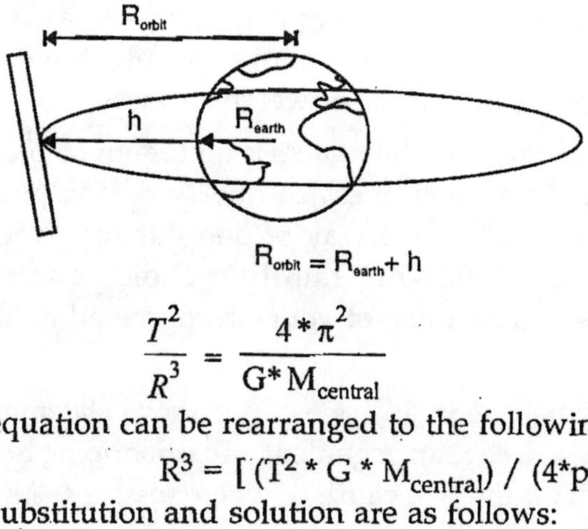

$$\frac{T^2}{R^3} = \frac{4*\pi^2}{G*M_{central}}$$

The equation can be rearranged to the following form

$$R^3 = [\,(T^2 * G * M_{central}) \,/\, (4*pi^2)\,]$$

The substitution and solution are as follows:

$$R^3 = [\,((86400\ s)^2 \bullet (6.673 \times 10^{-11}\ N\ m^2/kg^2) \bullet (5.98x10^{24}\ kg)) \,/\, (4 \bullet (3.1415)^2)\,]$$

$$R^3 = 7.54 \times 10^{22}\ m^3$$

By taking the cube root of 7.54 x 10^{22} m^3, the radius can be determined ro be

$$R = 4.23 \times 10^7\ m$$

The radius of orbit indicates the distance which the satellite is from the centre of the earth. Now that the radius of orbit has been found, the height above the earth can be calculated. Since the earth's surface is 6.37×10^6 m from its centre (that's the radius of the earth), the satellite must be a height of

$$4.23 \times 10^7 \text{ m} - 6.37 \times 10^6 \text{ m} = 3.59 \times 10^7 \text{ m}$$

above the surface of the earth. So the height of the satellite is 3.59×10^7 m.

WEIGHTLESSNESS IN ORBIT

Astronauts who are orbiting the Earth often experience sensations of weightlessness. These sensations experienced by orbiting astronauts are the same sensations experienced by anyone who has been temporarily suspended above the seat on an amusement park ride. Not only are the sensations the same (for astronauts and roller coaster riders), but the causes of those sensations of weightlessness are also the same. Unfortunately however, many people have difficulty understanding the causes of weightlessness.

The cause of weightlessness is quite simple to understand. However, the stubbornness of one's preconceptions on the topic often stand in the way of one's ability to understand. Consider the following multiple choice question about weightlessness as a test of your preconceived notions on the topic:

If you believe in any one of the above statements, then it might take a little rearrangement and remapping of your brain to understand the real cause of weightlessness. As is the case on many topics in Physics, some unlearning must first be done before doing the learning.

Put another way: it's not what you don't know which makes learning physics a difficult task; it's what you do know which makes learning physics a difficult task. So if you do have a preconception (or a strong preconception) about what weightlessness is, you need to be aware of that preconceived idea. And as you consider the following alternative conception about the meaning of weightlessness, evaluate the reasonableness and logic of the two competing ideas.

CONTACT VERSUS NON-CONTACT FORCES

Before understanding weightlessness, we will have to review two categories of forces - contact forces and action-at-a-distance forces. As you sit in a chair, you experience two forces - the force of the Earth's gravitational field pulling you downwards toward the Earth and the force of the chair pushing you upwards.

The upward chair force is sometimes referred to as a normal force and results from the contact between the chair top and your bottom end. This normal force is categorized as a contact force. Contact forces can only result from the actual touching of the two interacting objects - in this case, the chair and you. The force of gravity acting upon your body is not a contact force; it is often categorized as an action-at-a-distance force. The force of gravity is the result of your centre of mass and the Earth's centre of mass exerting a mutual pull on each other; this force would even exist if you were not in contact with the Earth.

The force of gravity does not require that the two interacting objects (your body and the Earth) make physical contact; it can act over a distance through space. Since the force of gravity is not a contact force, it cannot be felt through contact. You can never feel the force of gravity pulling upon your body in the same way that you would feel a contact force. If you slide across the asphalt tennis court (not recommended), you would feel the force of friction (a contact force). If you are pushed by a bully in the hallway, you would feel the applied force (a contact force).

If you swung from a rope in gym class, you would feel the tension force (a contact force). If you sit in your chair, you feel the normal force (a contact force). But if you are jumping on a trampoline, even while moving through the air, you do not feel the Earth pulling upon you with a force of gravity (an action-at-a-distance force).

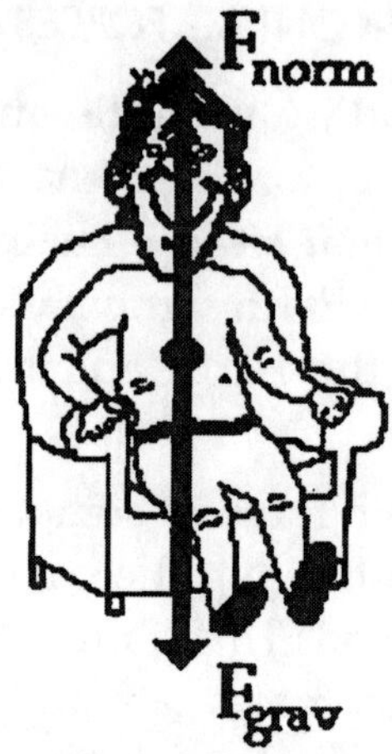

Fig. As you sit at rest in your chair, you feel the contact force (F_{norm}) balancing the non-contact force (F_{grav})

The force of gravity can never be felt. Yet those forces which result from contact can be felt. And in the case of sitting in your chair, you can feel the chair force; and it is this force which provides you with a sensation of weight. Since the upward normal force would equal the downward force of gravity when at rest, the strength of this normal force gives one a measure of the amount of gravitational pull. If there were no upward normal force acting upon your body, you would not have any sensation of your weight. Without the contact force (the normal force), there is no means of feeling the non-contact force (the force of gravity).

MEANING AND CAUSE OF WEIGHTLESSNESS

Weightlessness is simply a sensation experienced by an individual when there are no external objects touching one's body and exerting a push or pull upon it. Weightless sensations exist when all contact forces are removed. These sensations are common to any situation in which you are momentarily (or perpetually) in a state of free fall. When in free fall, the only force acting upon your body is the force of gravity - a non-contact force. Since the force of gravity cannot be felt without any other opposing forces, you would have no sensation of it. You would feel weightless when in a state of free fall.

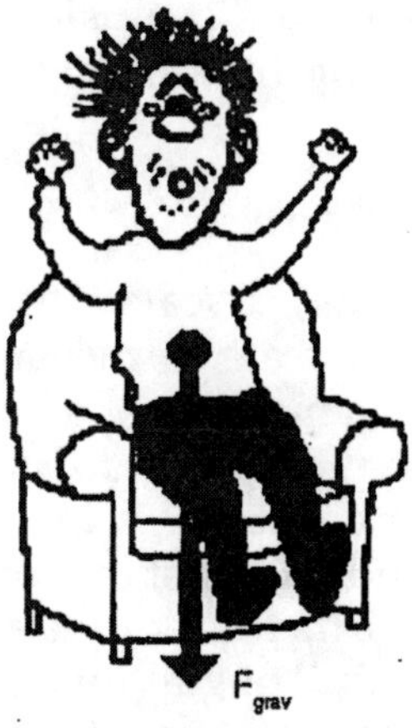

Fig. A person in free fall does not experience any contact force and thus feels weightless.

These feelings of weightlessness are common at amusement parks for riders of roller coasters and other rides in which riders are momentarily airborne and lifted out of their seats. Suppose that you were lifted in your chair to the top of a very high tower and then your chair was suddenly dropped. As you and your chair fall towards the ground, you both accelerate at the same rate - g. Since the chair is unstable, falling at the same rate as you, it is unable to push upon you. Normal forces only result from contact with stable, supporting surfaces. The force of gravity is the only force acting upon your body. There are no external objects touching your body and exerting a force. As such, you would experience a weightless sensation. You would weigh as much as you always do (or as little), yet you would not have any sensation of this weight.

Weightlessness is only a sensation; it is not a reality corresponding to an individual who has lost weight. As you are free-falling on a roller coaster ride (or other amusement park ride), you have not momentarily lost your weight. Weightlessness has very little to do with weight and mostly to do with the presence or absence of contact forces. If by "weight" we are referring to the force of gravitational attraction to the Earth, a free-falling person has not "lost their weight;" they are still experiencing the Earth's gravitational attraction. Unfortunately, the confusion of a person's actual weight with one's feeling of weight is the source of many misconceptions.

SCALE READINGS AND WEIGHT

Technically speaking, a scale does not measure one's weight. While we use a scale to measure one's weight, the scale reading is actually a measure of the upward force applied by the scale to balance the downward force of gravity acting upon an object. When an object is in a state of equilibrium (either at rest or in motion at constant speed), these two forces are balanced. The upwards force of the scale upon the person equals the downwards pull of gravity (also known as weight). And in this instance, the scale reading (which is a measure of the upwards force) equals the weight of the person. However, if you stand on the scale and bounce up and down, the scale reading undergoes a rapid change. As you undergo this bouncing motion, your body is accelerating. During the acceleration periods, the upward force of the scale is changing. And as such, the scale reading is changing. Is your weight changing? Absolutely not! You weigh as much (or as little) as you always do. The scale reading is changing, but remember: the Scale does not measure your weight. The scale is only measuring the external contact force which is being applied to your body.

Technically speaking...

... a scale does not measure your weight; a scale measures the upward force applied to your body.

Now consider Otis L. Evaderz who is conducting one of his famous elevator experiments. He stands on a bathroom scale and rides an elevator up and down. As he is accelerating upward and downward, the scale reading is different than when he is at rest and traveling at constant speed. When he is accelerating, the upward and downward force are not equal. But when he is at rest or moving at constant speed, the opposing forces balance each other. Knowing that the scale reading is a measure of the upward normal force of the scale upon his body, its value could be predicted for various stages of motion. For instance, the value of the normal force (F_{norm}) on Otis's 80-kg body could be predicted if the acceleration is known. This prediction can be made by simply applying Newton's second law. As an illustration of the use of Newton's second law to determine the varying contact forces on an elevator ride, consider the following diagram. In the diagram, Otis's 80-kg is traveling with constant speed (A), accelerating upward (B), accelerating downward (C), and free-falling (D) after the elevator cable snaps.

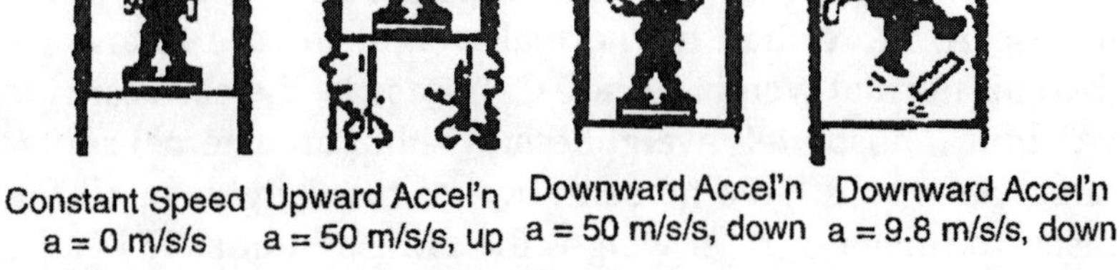

In each of these cases, the upward contact force (F_{norm}) can be determined using a free-body diagram and Newton's second law. The interaction of the two forces - the upwards normal force and the downwards force of gravity - can be thought of as a tug-of-war. The net force acting upon the person indicates who wins the tug-of-war (the up force or the down force) and by how much. A net force of 100-N, up indicates that the upwards force "wins" by an amount equal to 100 N. The gravitational force acting upon the rider is found using the equation $F_{grav} = m*g$. (An approximated g value of 10 m/s/s is used to simplify some of the mathematics and in turn make the conceptual principles more obvious.)

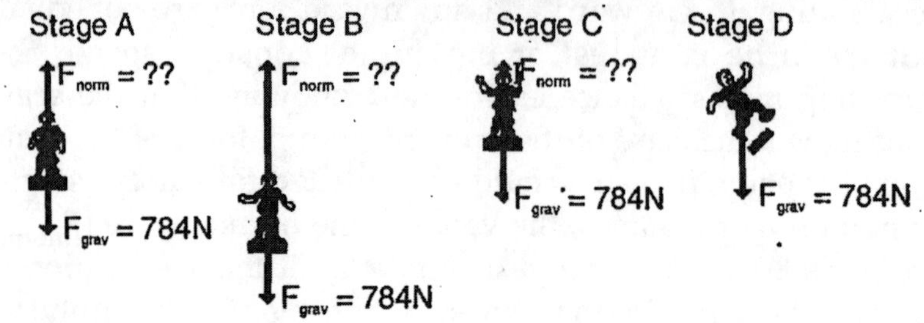

$F_{net} = m*a$	$F_{net} = m*a$	$F_{net} = m*a$	$F_{net} = m*a$
$F_{net} = 0$ N	$F_{net} = 400$ N, up	$F_{net} = 400$ N, down	$F_{net} = 784$ N, down
F_{norm} equals F_{grav}	$F_{norm} > F_{grav}$ by 400 N	$F_{norm} < F_{grav}$ by 400 N	$F_{norm} < F_{grav}$ by 784 N
$F_{norm} = 784$ N	$F_{norm} = 1184$ N	$F_{norm} = 384$ N	$F_{norm} = 0$ N

The normal force is greater than the force of gravity when there is an upwards acceleration (B), less than the force of gravity when there is a downwards acceleration (C and D), and equal to the force of gravity when there is no acceleration (A). Since it is the normal force which provides a sensation of one's weight, the elevator rider would feel his normal weight in case A, more than his normal weight in case B, and less than his normal weight in case C. In case D, the elevator rider would feel absolutely weightless; without an external contact force, he would have no sensation of his weight. In all four cases, the elevator rider weighs the same amount - 784 N. Yet the rider's sensation of his weight is fluctuating throughout the elevator ride.

WEIGHTLESSNESS IN ORBIT

Earth-orbiting astronauts are weightless for the same reasons that riders of a free-falling amusement park ride or a free-falling elevator are weightless. They are weightless because there is no external contact force pushing or pulling upon their body. In each case, gravity is the only force acting upon their body. Being an action-at-a-distance force, it cannot be felt and therefore would not provide any sensation of their weight. But for certain, the orbiting astronauts weigh something; that is, there is a force of gravity acting upon their body. In fact, if it were not for the force of gravity, the astronauts would not be orbiting in circular motion. It is the force of gravity which supplies the centripetal force requirement to allow the inward acceleration which is characteristic of circular motion. The force of gravity is the only force acting upon their body. The astronauts are in free-fall. Like the falling amusement park rider and the falling elevator rider, the astronauts and their surroundings are falling towards the Earth under the sole influence of gravity. The astronauts and all their surroundings - the space station with its contents - are falling towards the Earth without colliding into it. Their tangential velocity allows them to remain in orbital motion while the force of gravity pulls them inward.

Many students believe that orbiting astronauts are weightless because they do not experience a force of gravity. So to presume that the absence of gravity is the cause of the weightlessness experienced by orbiting astronauts would be in violation of circular motion principles. If a person believes that the absence of gravity is the cause of their weightlessness, then that person is hard-pressed to come up with a reason for why the astronauts are orbiting in the first place. The fact is that there must be a force of gravity in order for there to be an orbit.

One might respond to this discussion by adhering to a second misconception: the astronauts are weightless because the force of gravity is reduced in space. The reasoning goes as follows: "with less gravity, there would be less weight and thus they would feel less than their normal weight." While this is partly true, it does not explain their sense of weightlessness. The force of gravity acting upon an astronaut on the space station is certainly less than on Earth's surface. But how much less? Is it small enough to account for a significant reduction in weight? Absolutely not! If the space station orbits at an altitude of approximately 400 km above the Earth's surface, then the value of g at that location will be reduced from 9.8 m/s/s (at Earth's surface) to approximately 8.7 m/s/s. This would cause an astronaut weighing 1000 N at Earth's surface to be reduced in weight to approximately 890 N when in orbit. While this is certainly a reduction in weight, it does not account for the absolutely weightless sensations which astronauts experience. Their absolutely weightless sensations are the result of having "the floor pulled out from under them" (so to speak) as they are free-falling towards the Earth.

Still other physics students believe that weightlessness is due to the absence of air in space. Their misconception lies in the idea that there is no force of gravity when their is no air. According to them, gravity does not exist in a vacuum. But this is not the case. Gravity is a force which acts between the Earth's mass and the mass of other objects which surround it. The force of gravity can act across large distances and its affect can even penetrate across and into the vacuum of outer space. Perhaps students who own this misconception are confusing the force of gravity with air pressure. Air pressure is the result of surrounding air particles pressing upon the surface of an object in equal amounts from all directions. The force of gravity is not affected by air pressure. While air pressure reduces to zero in a location void of air (such as space), the force of gravity does not become 0 N. Indeed the presence of a vacuum results in the absence of air resistance; but this would not account for the weightless sensations. Astronauts merely feel weightless because there is no external contact force pushing or pulling upon their body. They are in a state of free fall.

ENERGY RELATIONSHIPS FOR SATELLITES

The orbits of satellites about a central massive body can be described as either circular or elliptical. As mentioned earlier in, a satellite orbiting about the earth in circular motion is moving with a constant speed and remains at the same height above the surface of the earth. It accomplishes this feat by moving with a tangential velocity that allows it to fall at the same rate at which the earth curves. At all instances during its trajectory, the force of gravity acts in a direction perpendicular to the direction which the satellite is moving. Since perpendicular components of motion are independent of each other, the inward force cannot affect the magnitude of the tangential velocity. For this reason, there is no acceleration in the tangential direction and the satellite remains in circular motion at a constant speed. A satellite orbiting the earth in elliptical motion will experience a component of force in the same or the opposite direction as its motion.

This force is capable of doing work upon the satellite. Thus, the force is capable of slowing down and speeding up the satellite. When the satellite moves away from the earth, there is a component of force in the opposite direction as its motion. During this portion of the satellite's trajectory, the force does negative work upon the satellite and slows it down. When the satellite moves towards the earth, there is a component of force in the same direction as its motion. During this portion of the satellite's trajectory, the force does positive work upon the satellite and speeds it up. Subsequently, the speed of a satellite in elliptical motion is constantly changing - increasing as it moves closer to the earth and decreasing as it moves further from the earth. These principles are depicted in the diagram below.

Satellite Motion-circular

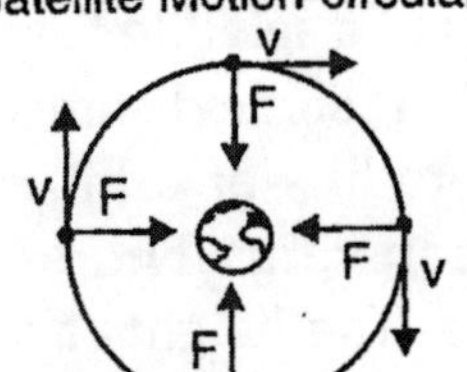

There is no component of force in the direction of motion.

Satellite Motion-Elliptical

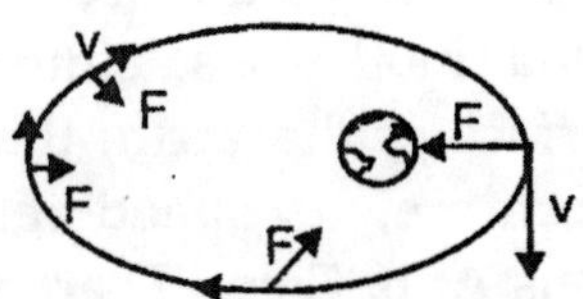

There is no component of force in the same and in the opposite direction as the object's motion.

Motion was analysed from an energy perspective. The governing principle which directed our analysis of motion was the work-energy theorem. Simply put, the theorem states that the initial amount of total mechanical energy (TME_i) of a system plus the work done by external forces (W_{ext}) on that system is equal to the final amount of total mechanical energy (TME_f) of the system. The mechanical energy can be either in the form of potential energy (energy of position - usually vertical height) or kinetic energy (energy of motion). The work-energy theorem is expressed in equation form as

$$KE_i + PE_i + W_{ext} = KE_f + PE_f$$

The W_{ext} term in this equation is representative of the amount of work done by external forces. For satellites, the only force is gravity. Since gravity is considered an internal (conservative) force, the W_{ext} term is zero. The equation can then be simplified to the following form.

$$KE_i + PE_i = KE_f + PE_f$$

In such a situation as this, we often say that the total mechanical energy of the system is conserved. That is, the sum of kinetic and potential energies is unchanging. While energy can be transformed from kinetic energy into potential energy, the total amount remains the same - mechanical energy is *conserved.* As a satellite orbits earth, its total mechanical energy remains the same. Whether in circular or elliptical motion, there are no external forces capable of altering its total energy.

ENERGY ANALYSIS OF CIRCULAR ORBITS

Let's consider the circular motion of a satellite first. When in circular motion, a satellite remains the same distance above the surface of the earth; that is, its radius of orbit is fixed. Furthermore, its speed remains constant. The speed at positions A, B, C and D are the same. The heights above the earth's surface at A, B, C and D are also the same. Since kinetic energy is dependent upon the speed of an object, the amount of kinetic energy will be constant throughout the satellite's motion. And since potential energy is dependent upon the height of an object, the amount of potential energy will be constant throughout the satellite's motion. So if the KE and the PE remain constant, it is quite reasonable to believe that the TME remains constant.

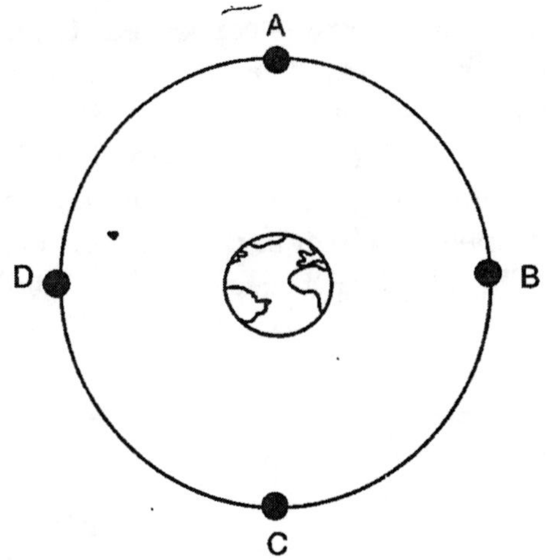

One means of representing the amount and the type of energy possessed by an object is a work-energy bar chart. A work-energy bar chart represents the energy of an object by means of a vertical bar. The length of the bar is representative of the amount of energy present - a longer bar representing a greater amount of energy. In a work-energy bar chart, a bar is constructed for each form of energy. A work-energy bar chart is presented below for a satellite in uniform circular motion about the earth. Observe that the bar chart depicts that the potential and kinetic energy of the satellite are the same at all four labeled positions of its trajectory (the diagram above shows the trajectory).

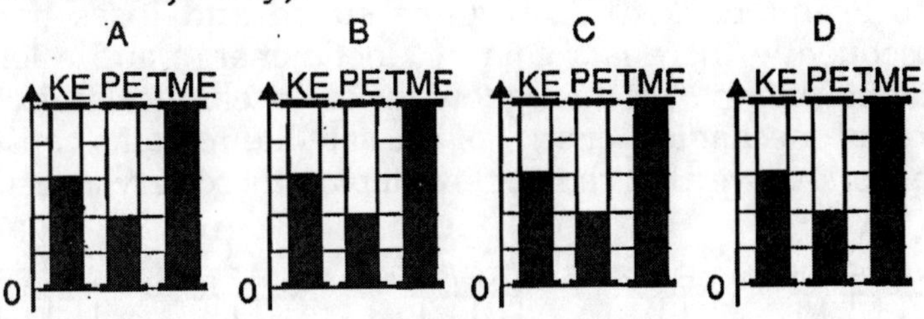

ENERGY ANALYSIS OF ELLIPTICAL ORBITS

Like the case of circular motion, the total amount of mechanical energy of a satellite in elliptical motion also remains constant. Since the only force doing work upon the satellite is an internal (conservative) force, the W_{ext} term is zero and mechanical energy is conserved. Unlike the case of circular motion, the energy of a satellite in elliptical motion will change forms. As mentioned above, the force of gravity does work upon a satellite to slow it down as it moves away from the earth and to speed it up as it moves towards the earth. So if the speed is changing, the kinetic energy will also be changing. The elliptical trajectory of a satellite is shown below.

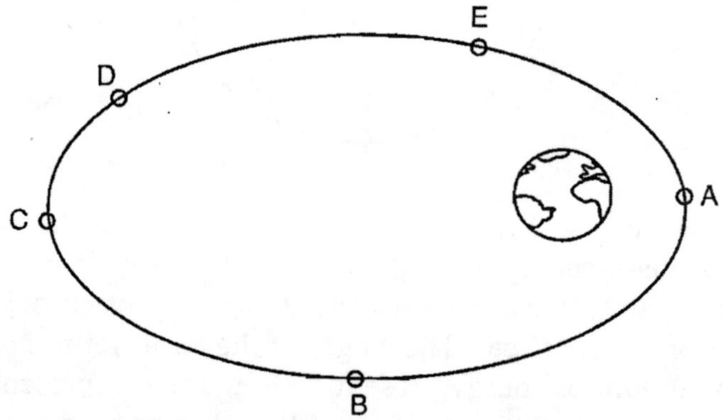

The speed of this satellite is greatest at location A (when the satellite is closest to the earth) and least at location C (when the satellite is furthest from the earth). So as the satellite moves from A to B to C, it loses kinetic energy and gains potential energy. The gain of potential energy as it moves from A to B to C is consistent with the fact that the satellite moves further from the surface of the earth. As the satellite moves from C to D to E and back to A, it gains speed and loses height; subsequently there is a gain of kinetic energy and a loss of potential energy. Yet throughout the entire elliptical trajectory, the total mechanical energy of the satellite remains constant. The work-energy bar chart below depicts these very principles.

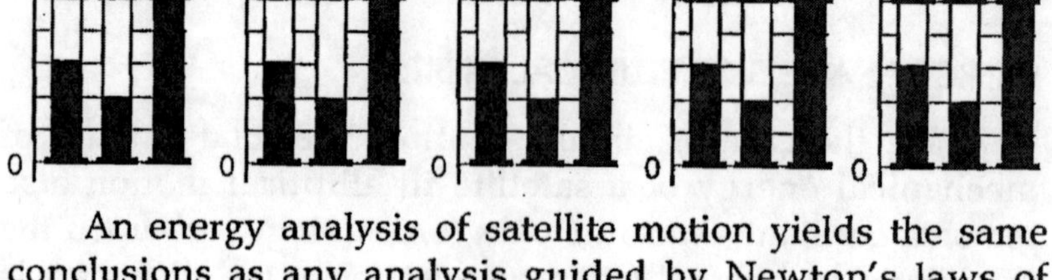

An energy analysis of satellite motion yields the same conclusions as any analysis guided by Newton's laws of motion. A satellite orbiting in circular motion maintains a constant radius of orbit and therefore a constant speed and a constant height above the earth. A satellite orbiting in elliptical motion will speed up as its height (or distance from the earth) is decreasing and slow down as its height (or distance from the earth) is increasing. The same principles of motion which apply to objects on earth - Newton's laws and the work-energy theorem - also govern the motion of satellites in the heavens.

NEWTON'S PRINCIPLES OF MECHANICS

Sir Isaac Newton realised that the force that makes apples fall to the ground is the same force that makes the planets "fall" around the sun. Newton had been asked to address the question of why planets move as they do. He established that a force of attraction toward the sun becomes weaker in proportion to the square of the distance from the sun.

Newton postulated that the shape of an orbit should be an ellipse. Circular orbits are merely a special case of an ellipse where the foci are coincident. Newton described his work in the *Mathematical Principles of Natural Philosophy* (often called simply the Principia), which he published in 1685. Newton gave his laws of motion as follows:

1. Every body continues in a state of rest, or of uniform motion in a straight line, unless it is compelled to change that state by forces impressed upon it.
2. The change of motion (linear momentum) is proportional to the force impressed and is made in the direction of the straight line in which that force is impressed.
3. To every action there is always an equal and opposite reaction; or, the mutual actions of two bodies upon each other are always equal, and act in opposite directions.

(Notice that Newton's laws describe the behaviour of inertia, they do not explain what the nature of inertia is. This is still a valid question, as is the nature of mass.)

There are three ways to modify the momentum of a body. The mass can be changed, the velocity can be changed (acceleration), or both.

Acceleration

Force (F) equals change in velocity (acceleration, A) times mass (M):

$$F = MA$$

Acceleration may be produced by applying a force to a mass (such as a spacecraft). If applied in the same direction as an object's velocity, the object's velocity increases in relation to an unaccelerated observer. If acceleration is produced by applying a force in the opposite direction from the object's original velocity, it will slow down relative to an unaccelerated observer. If the acceleration is produced by a force at some other angle to the velocity, the object will be deflected. These cases are illustrated below.

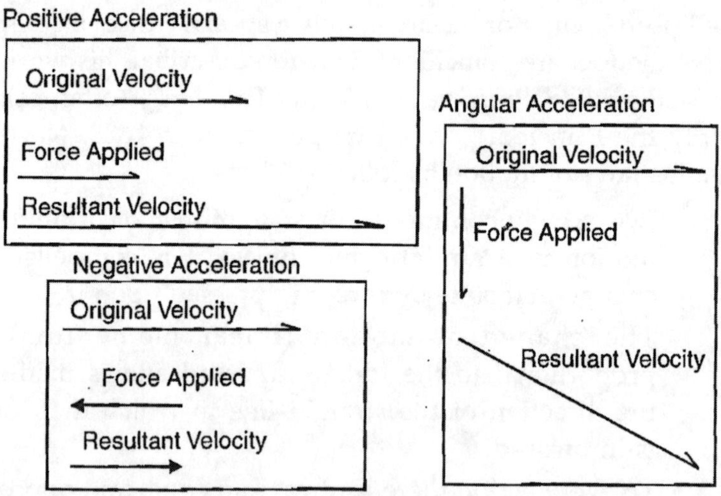

The world standard of mass is the kilogram, whose definition is based on the mass of a metal cylinder kept in France. Previously, the standard was based upon the mass of one cubic centimeter of water being one gram, which is approximately correct. The standard unit of force is the newton, which is the force required to accelerate a 1-kg mass 1 m/sec^2 (one meter per second per second). A newton is equal to the force from the weight of about 100 g of water in Earth's gravity. That's about half a cup. A dyne is the force required to accelerate a 1-g mass 1 cm/s^2.

UNDERSTANDING ACCELERATION

The relationship among velocity, acceleration and force is central to the study of mechanics. Text book treatment normally divides the scope of study between kinematics (study of velocity and acceleration) and dynamics (study of acceleration and force). This approach is perfectly fine till we do not encounter some inherent problem as to the understanding of the subject matter.

One important short coming of this approach is that we tend to assume that acceleration depends on the state of motion i.e. velocity. Such inference is not all too uncommon as acceleration is defined in terms of velocities. Now the question is "Does acceleration actually depend on the velocity?"

There are many other such inferences that need to be checked. Following paragraphs investigate these fundamental issues in detail and enrich our understanding of acceleration.

VELOCITY, ACCELERATION AND FORCE

Acceleration is related to external force. This relationship is given by Newton's second law of motion. For a constant mass system,

$$F = m\,a$$

In words, Newton's second law states that acceleration (effect) is the result of the application of net external force (cause). Thus, the relationship between the two quantities is that of cause and effect. Further, force is equal to the product of a scalar quantity, m, and a vector quantity, a, implying that the direction of the acceleration is same as that of the net force. This means that acceleration is though the measurement of the change of velocity, but is strictly determined by the external force and the mass of the body; and expressed in terms of "change of velocity" per unit time.

If we look around ourselves, we find that force modifies the state of motion of the objects. The immediate effect of a net force on a body is that the state of motion of the body changes. In other words, the velocity of the body changes in response to the application of net force. Here, the word "net" is important. The motion of the body responds to the net or resultant force. In this sense, acceleration is mere statement of the effect of the force as measurement of the rate of change of the velocity with time.

Now, there is complete freedom as to the magnitude and direction of force being applied. From our real time experience, we may substantiate this assertion. For example, we can deflect a foot ball, applying force as we wish (in both magnitude and direction). The motion of the ball has no bearing on how we apply force. Simply put: the magnitude and direction of the force (and that of acceleration) is not dependent on the magnitude and direction of the velocity of the body.

In the nutshell, we conclude that force and hence acceleration is independent of the velocity of the body. The magnitude and direction of the acceleration is determined by the magnitude and direction of the force and mass of the body. This is an important clarification.

To elucidate the assertion further, let us consider parabolic motion of a ball as shown in the figure. The important aspect of the parabolic motion is that the acceleration associated with motion is simply 'g' as there is no other force present except the force of gravity. The resultant force and mass of the ball together determine acceleration of the ball.

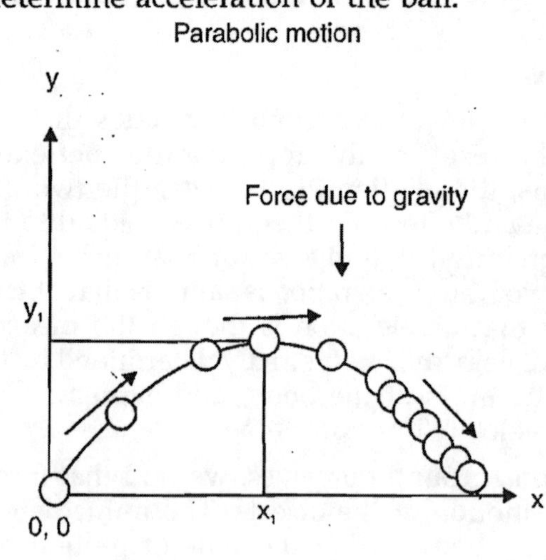

Example 1: Acceleration

Problem: If the tension in the string is T, when the string makes an angle θ with the vertical. Find the acceleration of a pendulum bob, having a mass "m".

Solution: As pointed out in our discussion, we need not study or consider velocities of bob to get the answer. Instead we should strive to know the resultant force to find out acceleration, using Newton's second law of motion.

The forces, acting on the bob, are

(i) Force of gravity, mg, acting in the downward direction and

(ii) Tension, T, acting along the string. Hence, the acceleration of the bob is determined by the resultant force, arising from the two forces.

Acceleration of a pendulum bob

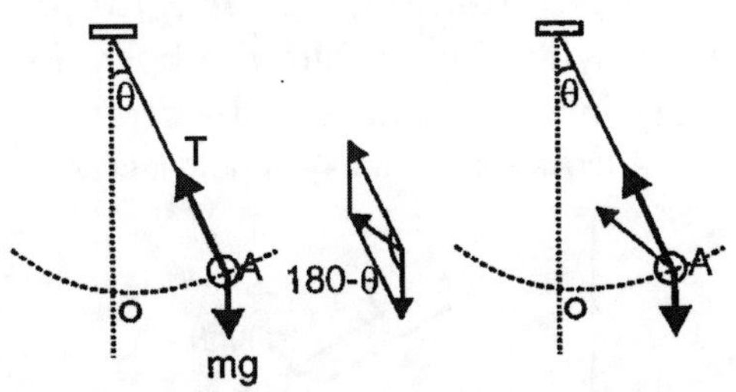

Using parallelogram theorem for vector addition, the resultant force is;

$$F = \sqrt{\left\{m^2 g^2 + T^2 + 2mgT\cos\left(180° - \theta\right)\right\}} =$$

$$\sqrt{\left(m^2 g^2 + T^2 + 2mgT\cos\right)}$$

The acceleration is in the direcion of force as shown in the figure, whereas the magnitude of the acceleration, a, is given by:

$$a = \sqrt{\frac{\left(m^2 g^2 + T^2 + 2mgT\cos\theta\right)}{m}}$$

External Force and Possible Scenarios

The change in velocity, resulting from the application of external force, may occur in magnitude or direction or both.

1. When force is applied in a given direction and the object is stationary.

   Under this situation, the magnitude of velocity increases with time, while the body follows a linear path in the direction of force (or acceleration).

2. When force is applied in the direction of the motion, then it increases the magnitude of the velocity without any change in the direction.

   Let us consider a block sliding on a smooth incline surface as shown in the figure. The component of the force due to gravity applies in the direction of motion. Under this situation, the magnitude of velocity increases with time, while the body follows a linear path. There is no change in the direction of motion.

A block sliding along smooth incline

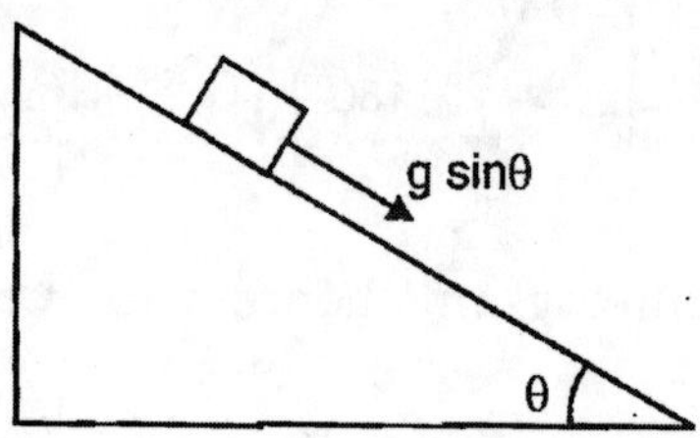

Fig. The block moves faster as it moves down the slope

3. When force is applied in the opposite direction to the motion, then it decreases the magnitude of the velocity without any change in the direction.

Take the example of a ball thrown vertically in the upward direction with certain velocity. Here, force due to gravity is acting downwards. The ball linearly rises to the maximum height till the velocity of ball reduces to zero. Then, the ball falls with increasing velocity as it reaches back to the initial position. In the adjoining figure, the vectors drawn at various points represent the direction and magnitude of force (F), acceleration (a) and velocity (v) during the motion. Note that both force and acceleration act in the downward direction during the motion.

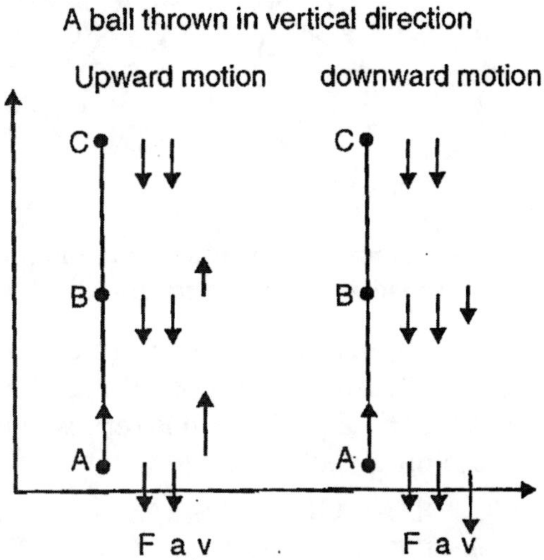

Fig. The ball reverses its direction during motion.

During upward motion, we see that the force acts in the opposite direction to that of the velocity. Under this situation, the magnitude of velocity decreases with time, while the body follows a linear path. There is no change in the direction of motion.

The velocity of the object at the point, where motion changes direction, is zero and force is acting downwards. This situation is same as the case 1. Hence, object moves in the direction of force i.e. in the downward direction.

4: When force is applied perpendicular to the direction of motion, then it causes change in the direction of the velocity.

A simple change of direction also constitutes change in velocity and, therefore, acceleration. Consider the case of a uniform circular motion in which a particle moves along a circular path at constant speed "v". Let V_1 & V_2 be the velocities of the particle at two time instants, then

$|v_1| = |v_2| =$ a constant

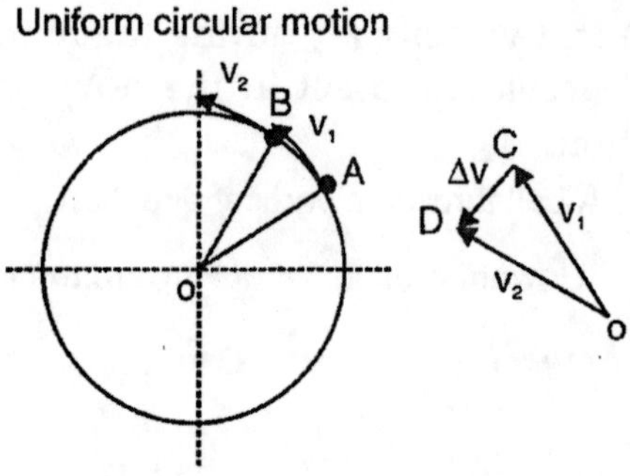

Fig. A central force perpendicular to motion causes chage in direction.

In the adjoining figure, the vector segments OC and OD represent V_1 & V_2. Knowing that vector difference Δv is directed from initial to final position, it is represented by vector CD. Using the adjoining vector triangle,

$$OC + CD = OD$$

$$v = CD = v_2 - v_1$$

The important thing to realise here is that direction of Δ v is along CD, which is directed towards the origin. This result is in complete agreement of what we know about uniform circular motion (The topic of uniform circular motion is covered in separate module). We need to apply a force (causing acceleration to the moving particle) across (i.e. perpendicular) to the motion to change direction. If the force (hence acceleration) is perpendicular to velocity, then magnitude of velocity i.e. speed remains same, whereas the direction of motion keeps changing.

Example 2: Acceleration

Problem: An object moves along a quadrant AB of a circle of radius 10 m with a constant speed 5 m/s. Find the average velocity and average acceleration in this interval.

Solution: Here, we can determine time interval form the first statement of the question. The particle covers a distance of 2πr/4 with a speed v. Hence, time interval,

$$t = \frac{\text{Distance}}{\text{Time}} = \frac{2\pi r}{4v} = \frac{3.14 \times 10}{2 \times 5} = 3.14s$$

Magnitude of average velocity is given by the ratio of the magnitude of displacement and time:

Motion along circular path

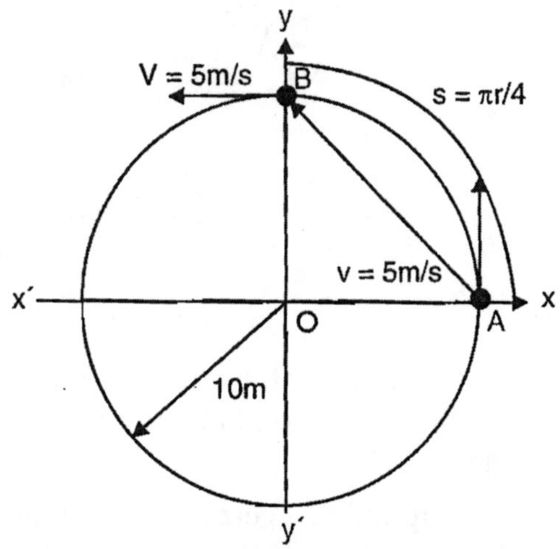

$$|V_{avg}| = \frac{|AB|}{Time} = \sqrt{\frac{(10^2+10^2)}{31.4}\,\frac{14.14}{3.14}} = 4.5\text{m/s}$$

Average velocity is directed along vector AB i.e. in the direction of displacement vector.

Motion along circular path

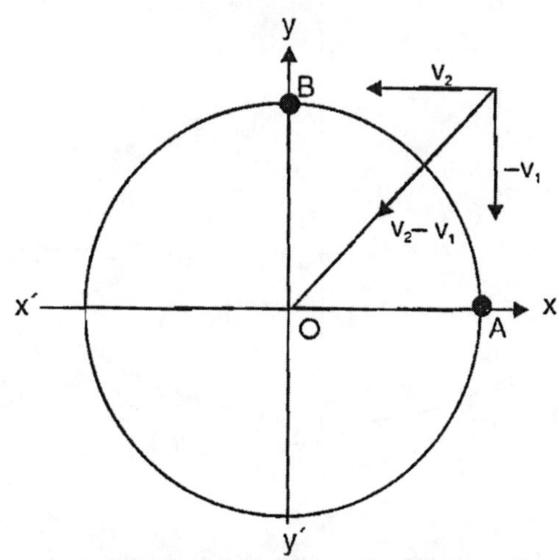

Fig. Direction of acceleration

Magnitude of average acceleration is given by the ratio:

Here,

$$a_{avg} = \frac{|\Delta v|}{\Delta t} \frac{\sqrt{(5^2 + 5^2)}}{3.14} \frac{5\sqrt{2}}{3.14} = 3.1 m/s^2$$

$$|\Delta v| = |v_2 - v_1|$$

$$|\Delta v| = \sqrt{(5^2 + 5^2)} = 5\sqrt{2} m/s$$

$$a_{avg} = \frac{5\sqrt{2}}{3.14} = 3.1 m/s^2$$

Average acceleration is directed along the direction of vector $V_1 - V_2$ i.e. directed towards the centre of the circle (as shown in the figure).

When force is applied at a certain angle with the motion, then it causes change in both magnitude and direction of the velocity. The component of force in the direction of motion changes its magnitude (an increase or decrease), while the component of force perpendicular to the direction of motion changes its direction.

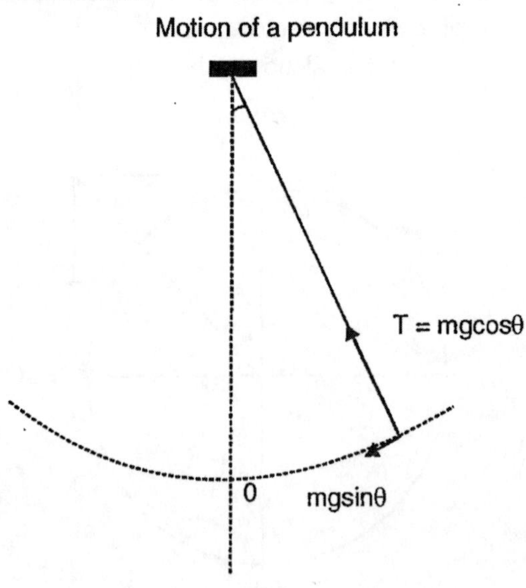

Fig. Forces acting on the pendulum bob

The motion of a small spherical mass "m", tied to a fixed point with the help of a string illustrates the changes taking place in both the direction and magnitude of the velocity. Saving the details for discussion at a later point in the course, here tension in the string and tangential force provide for the change in the direction and magnitude of velocity respectively.

When the string makes at angle, θ, with the vertical, then:

(a) Tension in the string, T is: $T = mg \cos\theta$. This force acts normal to the path of motion at all points and causes the body to continuously change its direction along the circular path. Note that this force acts perpendicular to the direction of velocity, which is along the tangent at any given point on the arc.

(b) Tangential force, F, is: $F = mg \sin\theta$. This force acts tangentially to the path and is in the direction of velocity of the spherical mass, which is also along the tangent at that point. As the force and velocity are in the same direction, this force changes the magnitude of velocity. The magnitude of velocity increases when force acts in the direction of velocity and magnitude of velocity decreases when force acts in the opposite direction to that of velocity.

EQUATIONS OF MOTION

In order to be accurate, the title of this section should be "One Dimensional Equations of Motion for Constant Acceleration". Given that such a title would be a stylistic nightmare, let me begin this section with the following qualification. The equations of motion are valid only when acceleration is constant and motion is constrained to a straight line.

Given that we live in a three dimensional universe in which the only constant is change, you may be tempted to dismiss this section outright. It would be correct to say that no object has ever traveled in a straight line with constant acceleration anywhere in the universe at any time – not today, not yesterday, not tomorrow, not five billion years ago, not thirty billion years in the future, never. This I can say with absolute metaphysical certainty.

So what good is this section then? Well, in many instances, it is useful to assume that an object did or will travel along a path that is essentially straight and with an acceleration that is nearly constant. That is, any deviation from the ideal motion can be essentially ignored. Motion along a curved path may also be effectively one-dimensional if there is only one degree of freedom for the objects involved. A road might twist and turn and explore all sorts of directions, but the cars driving on it have only one degree of freedom – the freedom to drive in one direction or the opposite direction. (You can't drive diagonally on a road and hope to stay on it for very long.) In this regard, it is not unlike motion restricted to a straight line. Approximating real situations with models based on ideal situations is not considered cheating. This is the way things get done in physics. It is such a useful technique that we will use it over and over again.

Our goal in this section, is to derive new equations that can be used to describe the motion of an object in terms of its three kinematic variables: velocity, displacement, and time. There are three ways to pair them up: velocity-time, displacement-time, and velocity-displacement. In this order, they are also often called the first, second, and third equations of motion, but there is no compelling reason to learn these names. Since we are dealing with motion in a straight line, the symbol x will be used for displacement. Direction will be indicated by the sign (positive quantities point in $+x$ direction, while negative quantities point in the ″x direction). Determining which direction is positive and which negative is entirely arbitrary. The laws of physics are isotropic; that is, they are independent of the orientation of the coordinate system. As long as you are consistent, it doesn't matter. Some problems are easier to understand and solve, however, when one direction is chosen positive over another.

Velocity-time

The relation between velocity and time is a simple one during constantly accelerated, straight-line motion. Constant acceleration implies a uniform rate of change in the velocity. The longer the acceleration, the greater the change in velocity. If after a time velocity increases by a certain amount, after twice that time it should increase by twice that amount. Change in velocity is directly proportional to time when acceleration is constant. If an object already started with a certain velocity, then its new velocity would be the old velocity plus this change. You ought to be able to see the equation in your mind's eye already.

This is the easiest of the three equations to derive formally. Start from the definition of acceleration, expand the Δv term, and solve for v as a function of t.

$$a = \frac{\Delta v}{\Delta t} = \frac{v - v_0}{\Delta t}$$

$$v = v_0 + a\Delta t$$

Since acceleration is also the first derivative of velocity with respect to time, this equation can also be derived using calculus. Just reverse the action of the definition. Instead of differentiating velocity to find acceleration, integrate acceleration to find velocity. Since acceleration is assumed constant, this is quite easy.

$$a = \frac{dv}{dt}$$

$$dv = adt$$

$$\int_{v_0}^{v} dv = \int_{0}^{\Delta t} adt$$

$$v - v_0 = a\Delta t$$

$$v = v_0 + a\Delta t$$

The symbol v_0 (v nought) is called the initial velocity. It is often thought of as the "first velocity" but this is a rather naïve way to describe it. Take the case of a meteor hurtling towards the earth. What is its initial velocity? If you want v_0 to be the first velocity, then the problem is over before it started. Who could possibly say what a meteor's velocity was at its birth?

There is no way to answer this question. A better definition would be to say that an initial velocity is the velocity that a moving object has when it first becomes important in a problem. Say the meteor was spotted deep in space and the problem was to determine its trajectory, then the initial velocity would be the velocity at the time it was observed. But if the problem were to determine its velocity on impact, then it's initial velocity would more likely be the velocity it had when it entered the earth's atmosphere. In this case, the answer to, "What's the initial velocity?" is "It depends". This turns out to be the answer to a lot of questions.

The symbol v is then the velocity some time Δt after the initial velocity. It is often called the final velocity but this does not make it an object's "last velocity". Take the case of the meteor. What velocity is represented by the symbol v? If you've been paying attention, then you should have anticipated the answer. It depends. It could be the velocity of the meteor as it passes by the moon, as it enters the earth's atmosphere, or as it strikes the earth's surface. It could also be the meteorite's velocity as it sits in the bottom of a crater (zero). Is this then the final velocity? Who knows. Someone could extract the meteorite from its hole in the ground and drive away with it. Is this relevant? Well, maybe. It depends. There's no rule for this kind of thing. You'll have to understand the problem and then make a decision for yourself.

The last part of this equation $a\Delta t$ is the change in the velocity from the initial value. Recall that a is the rate of change of velocity and that Δt is the time interval since the object had its initial velocity v_0. Rate multiplied by time equals change. Thus if an object were accelerating at 10 m/s^2, after 5 s it would be moving 50 m/s faster than it was initially. If it started with a velocity of 15 m/s, its velocity after 5 s of acceleration would be 65 m/s.

Displacement-time

The displacement of a moving object is directly proportional to both velocity and time. Move faster. Go farther. Move longer (as in longer time). Go farther. Acceleration compounds this simple situation. Now the velocity is also directly proportional to time. Try saying this in words and it sounds ridiculous. "Displacement is directly proportional to time and directly proportional to velocity, which is directly proportional to time." Time is a factor twice, making displacement proportional to the square of time. A car accelerating for two seconds would cover four times the distance of a car accelerating for only one second ($2^2 = 4$). A car accelerating for three seconds would cover nine times the distance ($3^2 = 9$).

Would that it should be so simple. This example only works when initial velocity is zero. Change in displacement is proportional to the square of time when acceleration is constant and initial velocity is zero. A true general statement would have to take into account any initial velocity and how the velocity was changing. This results in a terribly messy proportionality statement. Change in displacement is directly proportional to time and proportional to the square of time when acceleration is constant. A function that is both linear and square is said to be quadratic, which allows us to compact the previous statement considerably. Change in displacement is a quadratic function of time when acceleration is constant

Proportionality statements are useful, but not as concise as equations. We still don't know what the constants of proportionality are for this problem. The only way to answer that is through algebra.

Start with the definition of velocity, expand Δx, and solve it for displacement.

$$\bar{v} = \frac{\Delta x}{\Delta t}$$

$$\Delta x = \bar{v}\Delta t$$
$$x - x_0 = \bar{v}\Delta t$$
$$x = x_0 + \bar{v}\Delta t \qquad [a]$$

To continue, we need to resort to a little trick first published in the Fourteenth Century at Merton College, Oxford (and sometimes called the Merton Rule). When acceleration is constant, the velocity will change uniformly from its initial value to its final value and the average will lie halfway between the extremes. Thus, the average velocity is just the arithmetic mean of the initial and final velocities. Average velocity is the average of the final and initial velocities when acceleration is constant.

$$\bar{v} = \frac{v + v_0}{2}$$

Substitute the first equation of motion into this equation and simplify with the intent of eliminating v.

$$\bar{v} = \frac{v + v_0}{2} = \frac{(v_0 + a\Delta t) + v_0}{2} = \frac{2v_0 + a\Delta t}{2} = v_0 + \frac{1}{2}a\Delta t$$

Finally, substitute [b] into [a] and solve for x as a function of t.

$$x = x_0 + \bar{v}\Delta t$$
$$x = x_0 + (v_0 + a\Delta t)\Delta t$$
$$x = x_0 + v_0\Delta t + \frac{1}{2}a\Delta t^2$$

Since velocity is also the first derivative of displacement with respect to time, this equation can also be derived using calculus. In fact, it's much easier than using algebra. Just reverse the action of the definition. Instead of differentiating displacement to find velocity, integrate velocity to find displacement.

$$v = \frac{dx}{dt}$$
$$dx = v dt$$

$$\int_{x_0}^{x} dx = \int_{0}^{\Delta t} v dt = \int_{0}^{\Delta t} (v_0 + a\Delta t) dt$$

$$x - x_0 = v_0\Delta t + \frac{1}{2}a\Delta t^2$$

$$x = x_0 + v_0\Delta t + \frac{1}{2}a\Delta t^2$$

The symbol x_0 (ex nought) is the initial displacement. Many times, this value is zero and if it isn't, we can make it so. If you ask me, "When should we do this?" I would say, "It depends on the problem," and leave it to you to decide. There is no rule that you can memorize in this case. You have to understand what the equation says and then learn how to apply it to a particular situation. Similarly x is often called the final displacement, but this does not make it the "last displacement", rather it is the displacement at the end of the time interval during which the acceleration was constant.

Something else to notice is the similarity between equations. When acceleration is zero, our second equation of motion reverts to a rearranged constant velocity equation. As predicted, displacement is in part directly proportional to time and in part directly proportional to time squared.

$$x = x_0 + v_0\Delta t + \frac{1}{2}a\Delta t^2$$

$$x = x_0 + \bar{v}\Delta t$$

Although the velocity symbols in the two equations may look different, they do indeed represent the same quantity. If there is no acceleration, then the velocity is constant, which means that the initial velocity is the same as the final velocity is the same as the average velocity. The acceleration term at the end is an adjustment to the constant velocity equation to account for the the fact that the velocity is changing. A positive acceleration would increase the displacement and a negative acceleration would decrease it. This is exactly what one would expect. If an object's velocity was increasing, it would move farther than if it had stayed at a constant velocity. Likewise, if an object's velocity was decreasing, it would have a smaller displacement than if its velocity were constant. It's good to see that the equations behave in a realistic manner. Otherwise all this math would be a waste of time.

Velocity-displacement

We have just seen that velocity is directly proportional to time and displacement is proportional to time squared. With a little bit of thinking, a new proportionality statement should be apparent. Change in displacement is proportional to the change in the square of velocity when acceleration is constant. This statement is particularly important for driving safety. When you double the speed of a car, it takes four times the distance to stop it. Triple the speed and you'll need nine times the distance. Like the previous relationships, it also depends on initial velocity. Unfortunately determining the effect using reasoning alone is a real chore. Do the algebra instead.

The last two equations each described one kinematic variable as a function of time. It would be nice if we also had an equation that was independent of time. That is, we want to answer the question, "What is the relationship between velocity and displacement?" The method of doing this should be readily apparent. We've got to combine our first two equations of motion together in a manner that will eliminate time as a variable. The easiest way to do that is to solve one equation for time and then substitute it into the other. The second equation of motion is a quadratic and solving it for time would introduce a lot of nastiness into the algebra. It should be apparent that solving the first equation of motion for time and substituting it into the second will be the easier process.

$$x = x_0 + v_0\Delta t + \frac{1}{2}a\Delta t^2$$

$$v = v_0 + a\Delta t$$

$$x - x_0 = v_0\Delta t + \frac{1}{2}a\Delta t^2 \Leftarrow \Delta t = \frac{v - v_0}{a}$$

$$x - x_0 = v_0\frac{(v - v_0)}{(a)} + \frac{1}{2}a\frac{(v - v_0)^2}{(a)}$$

$$x - x_0 = \frac{vv_0 - v_0^2}{a} + \frac{v^2 - 2vv_0 + v_0^2}{2a}$$

$$2a(x - x_0) = \left(2vv_0 - 2v_0^2\right) + \left(v^2 - 2vv_0 + v_0^2\right)$$

$$2a(x - x_0) = v^2 - v_0^2$$

$$v^2 = v_0^2 + 2a(x - x_0)$$

As expected, displacement is proportional to velocity squared. (Initial velocity squared is just a constant that must be dealt with.) Unlike the first and second equations of motion, there is no advantage in using calculus to derive the third equation of motion. Algebra is the way to go.

Summary

- The equations of motion are valid only when ...
  - Acceleration is constant and
  - Motion is constrained to a straight line.

Table: The one Dimensional Equations of Motion for Constant Acceleration

Traditional Name	Equation	Relationship
1st equation	$v = v_0 + at$	velocity - time
2nd equation	$x = x_0 + v_0 t + \frac{1}{2}at^2$	displacement - time
3rd equation	$v^2 = v_0^2 + 2a(x - x_0)$	velocity - displacement
merton rule	$\bar{v} = \frac{1}{2}(v + v_0)$	average velocity

NON-NEWTONIAN PHYSICS

We know from Einstein's special theory of relativity that mass, time, and length are variable and the speed of light is constant. And from general relativity, we know that gravitation and acceleration are equivalent, that light bends in the presence of mass, and that an accelerating mass radiates gravitational waves at the speed of light.

Spacecraft operate at very high velocities compared to velocities we are familiar with in transportation and ballistics here on our planet. Since spacecraft velocities do not approach a significant fraction of the speed of light, Newtonian physics serves well for operating and navigating throughout the solar system. Nevertheless, accuracies are routinely enhanced by accounting for tiny relativistic effects. Once we begin to travel between the stars, velocities may be large enough fractions of light speed that Einsteinian physics will be indespensible for determining trajectories.

For now, spacecraft do sometimes carry out experiments to test special relativity effects on moving clocks, and experiments to test general relativity effects such as the space-time warp caused by the sun, frame-dragging, the equivalence of acceleration and gravitation (more precisely the equivalence between inertial mass and gravitational mass) and the search for direct evidence of gravitational waves. As of April 2006 there has been no test by which an observer could tell acceleration from gravitation, nor has gravitational radiation been directly observed

FIRST PRINCIPLES OF MECHANICS.

Equilibrium.—When two or more forces operate on a body in opposite directions, so as to counteract or balance each other, they produce what is called equilibrium. If two horses of equal power are harnessed to a reed, or a small slender tree, and draw in opposite directions, the tree will not be bent by the draught. If four horses draw at the same object in the direction of the east, west, north, south, no effect will be thereby produced. The exertion of forces in opposite vertical direction, are equally ineffectual. In the diagram which heads this article, the hoop A. is supposed to be void of weight, or as light as air; but has the force of the weight of several equal balls drawing in different directions.

The consequence is, that the hoop remains unmoved. But let either of the cords which sustain the balls be severed, and the hoop would leave its present position instantly. In this illustration, the only force applied to the hoop is that of gravity. If the hoop was made of iron, and was equal in weight to two of the balls, and the two lower balls were detached, the hoop would still remain unmoved; being then as now represented, in a state of perfect equilibrium. Thus a stone or any other ponderous body, while lying at rest on the surface of the earth, is subject to the action or exertion of the force of gravity in a variety of directions, yet it has no tendency to move in any direction except downward. When a block of wood is gently put into still water, it will descend by the force of gravity, until it has displaced a quantity of water, equal to its own weight.

If an inch cube, of any substance whose specific gravity is equal to that of water, be put into a vessel of water containing two square inches of surface, the cube will descend half an inch, and in so doing, it will elevate the surface of the water half an inch, thus bringing the surface of the water to a level with the top of the cube.

In this instance it may be readily understood, that by the descent of the cube in the water, an equal quantity and weight of water is elevated a distance equal to that which the cube descends, thus seeking and producing an equality of force. In this instance the water exerts a force termed buoyancy on the cube and directly opposite to that of gravity. In all instances of the action of buoyancy, in which any ponderous body is supported by water, the whole surface of the water, however extensive that surface may be, is elevated: the extent and elevation of this always bears the same proportion to the descent of the floating body, that the area of the floating body does to the whole surface of the water.

An interesting instance of equilibrium, may sometimes be seen in a square rigged sailing vessel, directly before the wind, but stemming a strong current. Cases of this kind, may be often seen on the Connecticut River, in which, vessels nearly remain stationary for Hours, having a strong current in one direction, and fresh breeze in the opposite, while gravity direct—the weight of the vessel and cargo—tends to sink or depress it, and the force of buoyancy of the water supports or holds it up; thus the vessel is held in perfect equilibrium. We have not supposed it necessary to give any explanation of the nature of the pulleys, over which the cords pass from the hoop to the balls, as represented above; but as the lever, of which the pulley is a modification, constitutes an important item in the science of mechanics, we shall make that the subject of the Scientific Mechanic.

SIMPLE HARMONIC MOTION

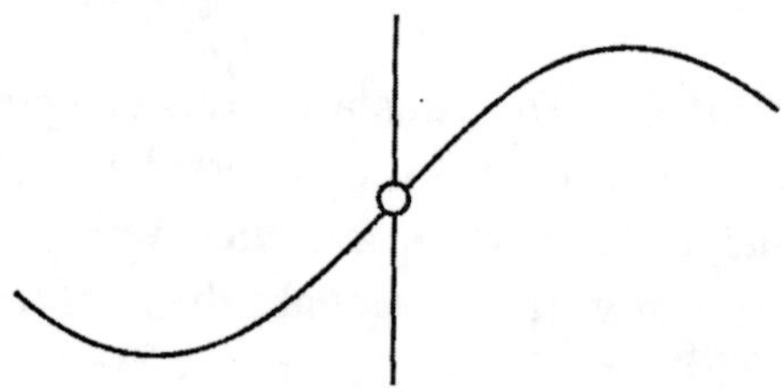

Fig. Simple harmonic motion.

Simple harmonic motion is the motion of a simple harmonic oscillator, a motion that is neither driven nor damped. The motion is periodic, as it repeats itself at standard intervals in a specific manner - described as being sinusoidal, with constant amplitude. It is characterized by its amplitude (which is always positive), its period which is the time for a single oscillation, its frequency which is the number of cycles per unit time, and its phase, which determines the starting point on the sine wave. The period, and its inverse the frequency, are constants determined by the overall system, while the amplitude and phase are determined by the initial conditions (position and velocity) of that system.

Overview

Simple harmonic motion is defined by the differential equation, $\frac{d^2x}{dt^2} = -kx$, where k is a positive constant.

In words simple harmonic motion is "motion where the acceleration of a body is proportional to, and opposite in direction to the displacement from its equilibrium position".

The above differentional equation has a sinusoidal solution, pictured below.

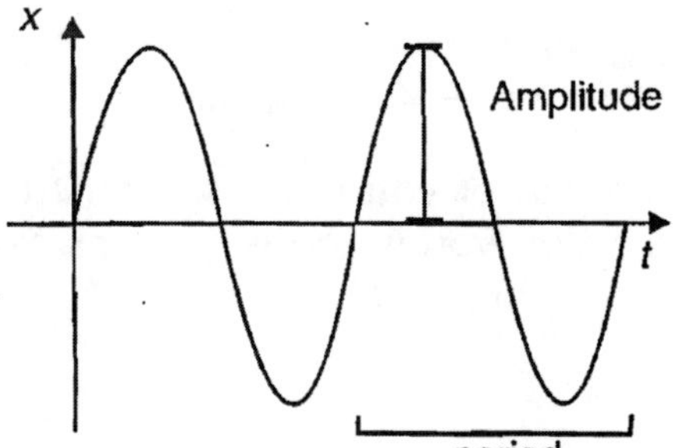

A general equation describing simple harmonic motion is $x(t) = A\sin(2\pi ft + \delta)$, where x is the displacement, A is the amplitude of oscillation, f is the frequency, t is the elapsed time, and Δ is the phase of oscillation. If there is no displacement at time t = 0, the phase Δ = 0. A motion with frequency f has period.

$$T = \frac{1}{f}$$

Simple harmonic motion can serve as a mathematical model of a variety of motions and provides the basis of the characterisation of more complicated motions through the techniques of Fourier analysis.

MATHEMATICS

It can be shown, by differentiating, exactly how the acceleration varies with time. Using the angular frequency ω, defined as

$$\omega = 2\pi f$$

the displacement is given by the function

$$x(t) = A\sin(\omega t + \delta)$$

Differentiating once gives an expression for the velocity at any time.

$$v(t) = \frac{dx(t)}{dt} = A\omega\cos(\omega t + \delta)$$

And once again to get the acceleration at a given time.

$$a(t)=\frac{d^2x(t)}{dt^2} = -A\omega^2 \sin(\omega t+\delta)$$

These results can of course be simplified, giving us an expression for acceleration in terms of displacement.

$$a(t) = \omega^2 x(t).$$

Examples

An undamped spring-mass system undergoes simple harmonic motion.

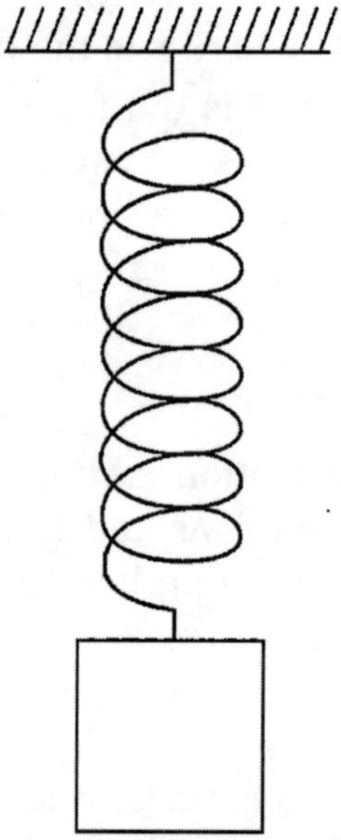

Simple harmonic motion is exhibited in a variety of simple physical systems and below are some examples:

Mass on a spring: A mass M attached to a spring of spring constant k exhibits simple harmonic motion in space with

$$\omega = 2\pi f = \sqrt{\frac{k}{M}}$$

Alternately, if the other factors are known and the period is to be found, this equation can be used:

$$T = \frac{1}{f} = 2\pi\sqrt{\frac{M}{k}}$$

The total energy is constant, and given by $E = \frac{kA^2}{2}$ where E is the total energy.

Uniform circular motion: Simple harmonic motion can in some cases be considered to be the one-dimensional projection of uniform circular motion. If an object moves with angular frequency ω around a circle of radius R centreed at the origin of the x-y plane, then its motion along the x and the y coordinates is simple harmonic with amplitude R and angular speed ω.

Mass on a pendulum: In the small-angle approximation, the motion of a pendulum is approximated by simple harmonic motion. The period of a mass attached to a string of length l with gravitational acceleration g is given by

$$T = 2\pi\sqrt{\frac{l}{g}}$$

This approximation is accurate only in small angles because of the expression for angular acceleration being proportional to the sine of position:

$$lmg \sin g(\theta) = I\alpha$$

where I is the moment of inertia; in this case $I = ml^2$. When θ is small, $\sin(\theta) \approx \theta$ and therefore the expression becomes

$$\text{Img}\theta = \text{I}\alpha$$

which makes angular acceleration directly proportional to θ, satisfying the definition of Simple Harmonic Motion.

For a solution not relying on a small-angle approximation, see pendulum (mathematics).

UNIFORM CIRCULAR MOTION

There are two types of circular motion: uniform circular motion and non-uniform circular motion.

Uniform circular motion describes motion in which an object moves with constant speed along a circular path.

Acceleration and Velocity

Since the velocity is tangent to the circular path, no two velocities point in the same direction. Although the object has a constant *speed*, its *direction* is always changing. This change in velocity is caused by an acceleration, whose magnitude is (like that of the velocity) held constant, but whose direction is always changing. The acceleration points radially inwards (centripetally) and is perpendicular to its velocity. This acceleration is known as centripetal acceleration.

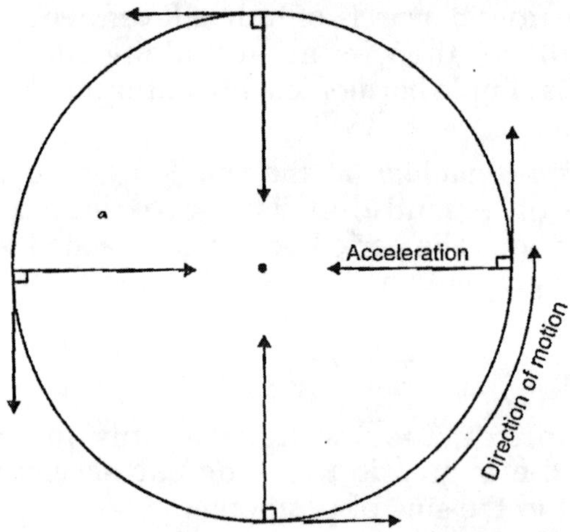

The magnitude of the acceleration is given by $a = v^2 / r$, where v is the speed of the object and r is the radius of its path, or $a = (4\pi^2 r) / t^2$, where r is the radius of the object and t is the time it takes the object to travel a distance.

DERIVATION OF THE EQUATIONS OF ACCELERATION

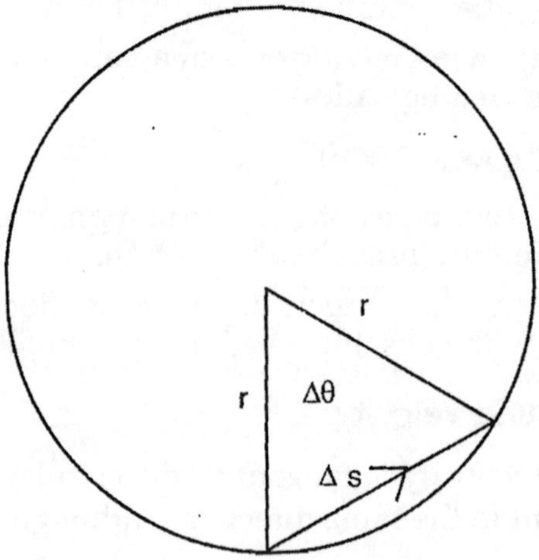

The above picture shows a point of mass that is moving with a constant angular speed around a centre. When the change in angle is Δθ, the change in displacement is Δs. Using the relationship of trigonometric functions, we find that,

$$\Delta s = \frac{r \sin \Delta\theta}{\cos(\Delta\theta/2)}$$

This equation is only valid when Δθ does not equal $(2n + 1)\pi$ where n is integer.

Similarly, the magnitude of tangential speed is always the same. Let Δv be the change in velocity, v be the initial velocity or instantaneous velocity, and Δt be the change in time,

$$\Delta v = \frac{v \sin \Delta\theta}{\cos(\Delta\theta/2)}$$

$$\frac{\Delta s}{\Delta t} = \frac{r \sin \Delta\theta}{\Delta t \cos(\Delta\theta/2)}$$

When $\Delta t \to 0, \Delta\theta \to 0,$

$$\lim_{\Delta t \to 0} \frac{\Delta s}{\Delta t} = \lim_{\Delta t \to 0} \frac{r \sin \Delta\theta}{\Delta t \cos(\Delta\theta/2)}$$

$$\text{v} = \lim_{\Delta t \to 0} \frac{r\Delta\theta}{\Delta t}$$

$v = r\frac{d\theta}{dt} = rw$ (ω is angular speed)

$$\frac{r}{\Delta s} = \frac{\cos(\Delta\theta/2)}{\sin \Delta\theta} = \frac{v}{\Delta v}$$

$$\Delta v = \frac{v\Delta s}{r}$$

$$\frac{\Delta v}{\Delta t} = \frac{v\Delta s}{r\Delta t}$$

$$\lim_{\Delta t \to 0} \frac{\Delta v}{\Delta t} = \lim_{\Delta t \to 0} \frac{v\Delta s}{r\Delta t}$$

$$a = \frac{v^2}{r}$$

Because $v = r\omega$

We can substitute into $a = \frac{v^2}{r}$

to get: $a = \frac{(r\omega)^2}{r} = r\omega^2$

CENTRIPETAL FORCE

The acceleration is usually considered to be due to an inward acting force, which is known as the Centripetal force. Centripetal force means "centre seeking" force. It is the force that keeps an object in its uniform circular motion. We determine this force by using Newton's second law of motion, $F_{net} = ma$, where F_{net} is the net force acting on the object (this is the centripetal force, F_c, of an object in uniform circular motion), m is the mass of the object, and a is the acceleration of the object. Since the acceleration of the object in uniform circular motion is the centripetal acceleration, we can substitute v^2 / r or $(4\pi^2 r) / t^2$ for a. This gets us $F_c = (mv^2) / r$ or $F_c = (4m\pi^2 r) / t^2$

The centripetal force can be provided by many different things, such as tension (as in a string), and friction (as between a tire and the road).

An example of tension being the centripetal force is tying a mass onto a string and spinning it around in a horizontal circle above your head. The tension force is the centripetal force because it is the only force keeping the object in uniform circular motion.

The m is the mass of the object, and the tension force is the centripetal force because it is keeping the object in uniform circular motion.

If a person were to cut the rope at one given point, the object would continue to move in the direction of the velocity, tangent to the circular path.

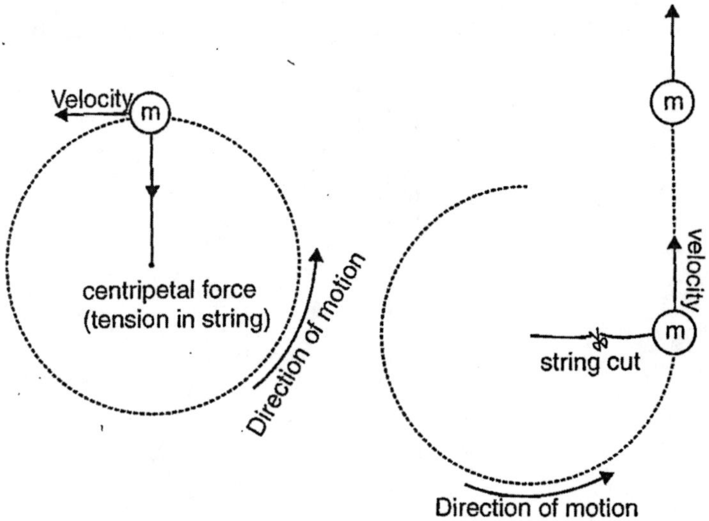

As one can see, the string holding mass *m* is cut about ¾ of the way. After the string is cut, the tension force/centripetal force is no longer acting upon the object so there is no force holding the object in uniform circular motion. Therefore it continues going in the direction when it was last in contact with the force.

Similar to the tension force, the friction force between the tires of a car and the road is the centripetal force because it keeps the car moving in a circular path. If this were a frictionless plane, the car would not be able to move in uniform circular motion, and will instead travel in a straight line. Without the friction force acting upon it, no force is keeping the car in uniform circular, so it moves in a straight line.

PENDULUM

The mathematics of pendulums can be quite complex, but some formulas and proofs are given below.

Simple Gravity Pendulum

A simple pendulum is an ideality involving these two assumptions:

- The rod/string/cable on which the bob is swinging is massless and always remains taut;

- Motion occurs in a 2-dimensional plane, i.e. the bob does not trace an ellipse.

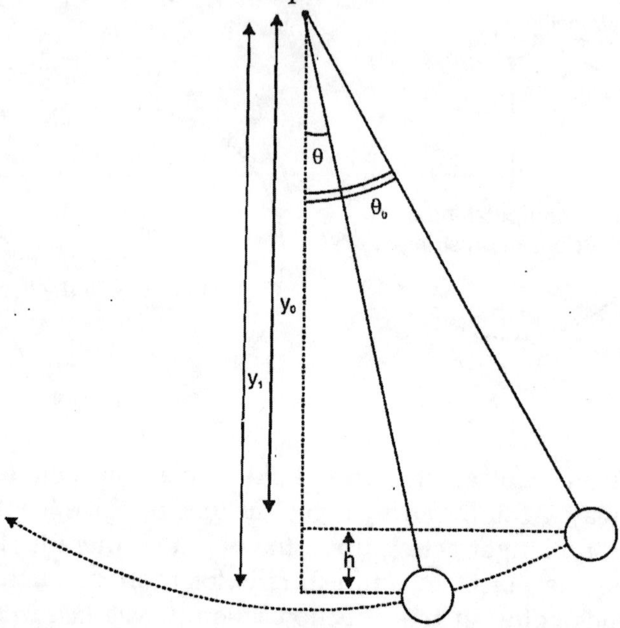

Fig. Trigonometry of a simple gravity pendulum.

The differential equation which represents the approximate motion of the pendulum is

$$\frac{d^2\theta}{dt^2}+\frac{g}{l}\sin\theta = 0$$

This is known as Mathieu's equation. It can also be obtained via the conservation of mechanical energy principle: any given object which fell a vertical distance h would have acquired kinetic energy equal to that which it lost to the fall. In other words, gravitational potential energy is converted into kinetic energy.

The *first integral of motion* is

$$\frac{d\theta}{dt} = \sqrt{\frac{2g}{l}(\cos\theta-\cos\theta_0)}$$

It gives the velocity in terms of the location and includes an integration constant related to the initial displacement (θ_0).

SMALL-ANGLE APPROXIMATION

The problem with the equations developed in the previous section is that they are unintegrable. To shed some light on the behaviour of the pendulum we shall make another approximation. Namely, we restrict the motion of the pendulum to a relatively small amplitude, that is, relatively small θ. How small? Small enough that the following approximation is true within some desirable tolerance

$\sin\theta \approx \theta$

if and only if

$|\theta| \ll 1$

Substituting this approximation into yields

$$\frac{d^2\theta}{dt^2} + \frac{g}{l}\theta = 0$$

Under the initial conditions $\theta(0) = \theta_0$ and $\frac{d\theta}{dt}(0) = 0$, the solution to this equation is a well-known, and quite expected, oscillatory function

$$\theta(t) = \theta_0 \cos\left(\sqrt{\frac{g}{l}}t\right) \qquad |\theta_0 \ll 1|$$

where θ_0 is the semi-amplitude of the oscillation, that is the maximum angle between the rod of the pendulum and the vertical.

The term $\sqrt{\frac{g}{l}}$ is a pulsation, which is equal to $\frac{2\pi}{T_0}$,

where T_0 is the period of a complete oscillation (outward and return).

$$\text{Since } w = \sqrt{\frac{g}{l}} = \frac{2\pi}{T_0}$$

$$T_0 = 2\pi\sqrt{\frac{l}{g}} \qquad |\theta_0| \ll 1$$

the period of a complete oscillation can be easily found, and we have obtained Huygens's law:

Further approximation

$$T_0 = 2\pi\sqrt{\frac{l}{g}}$$

can be expressed as $l = \frac{g}{\pi^2} \times \frac{T_0^2}{4}$

If we use SI units (i.e. measure in metres and seconds), and assume the measurement is taking place on the earth's surface, then $g = 9.80665$ m/s², and $f\frac{g}{\pi^2} \approx 1$ (the exact figure is 0.994 to 3 decimal places).

Therefore $l \approx \frac{T_0^2}{4}$ or to put it in words:

On the surface of the earth, the length of a pendulum (in metres) is approximately one quarter of the time period (in seconds) squared.

Arbitrary-amplitude Period

For amplitudes beyond the small angle approximation, one can compute the exact period by inverting equation

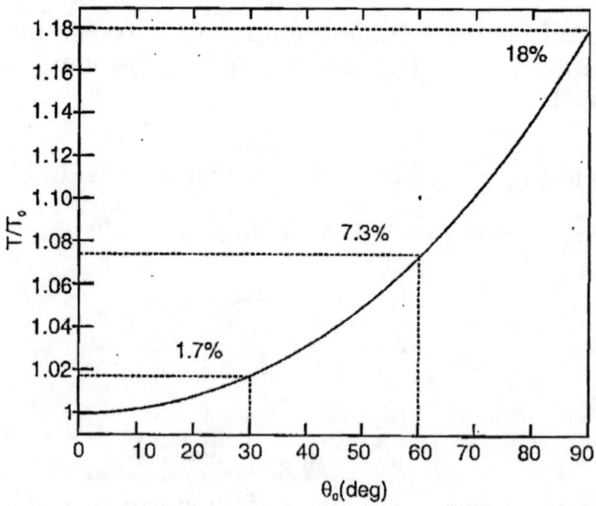

Fig. Deviation of the period from small-angle approximation.

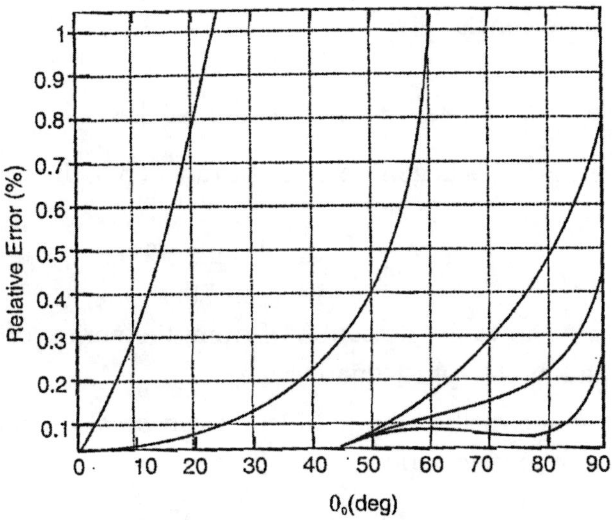

Fig. Relative errors using the power series.

$$\frac{dt}{d\theta} = \frac{1}{\sqrt{2}}\sqrt{\frac{l}{g}\frac{1}{\sqrt{\cos\theta - \cos\theta_0}}}$$

and integrating over one complete cycle,

$$T = \theta_0 \rightarrow 0 \rightarrow -\theta_0 \rightarrow 0 \rightarrow \theta_0,$$

or twice the half-cycle

$$T = 2(\theta_0 \rightarrow 0 \rightarrow -\theta_0)$$

or 4 times the quarter-cycle

$$T = 4(\theta_0 \rightarrow 0)$$

which leads to

$$T = 4\frac{1}{\sqrt{2}}\sqrt{\frac{l}{g}}\int_0^{\theta_0}\frac{1}{\sqrt{\cos\theta - \cos\theta_0}}d\theta$$

Alas, this integral cannot be evaluated in terms of elementary functions. It can be re-written in the form of the elliptic function of the first kind, which gives one very little advantage for it is a redundant exercise of expressing one insoluble integral in terms of another

$$T = 4\sqrt{\frac{l}{g}}F\left(\frac{\theta_0}{2}, \csc^2\frac{\theta_0}{2}\right)\csc\frac{\theta_0}{2}$$

or more concisely,

$$T = 4\sqrt{\frac{l}{g}}F\left(\sin\frac{\theta_0}{2},\frac{\pi}{2}\right)$$

where $F(k,ö)$ is Legendre's elliptic function of the first kind

$$F(k,\phi) = \int_0^\phi \frac{1}{\sqrt{1-k^2\sin^2\theta}}d\theta$$

Figure shows the deviation of from; the period obtained from small-angle approximation.

The value of the elliptic function can be also computed using the following series:

$$T = 2\pi\sqrt{\frac{l}{g}}\left(1+\left(\frac{1}{2}\right)^2\sin^2\left(\frac{\theta_0}{2}\right)+\right.$$

$$\left.\left(\frac{1\cdot3}{2\cdot4}\right)^2\sin^4\left(\frac{\theta_0}{2}\right)+\left(\frac{1\cdot3\cdot5}{2\cdot4\cdot6}\right)^2\sin^6\left(\frac{\theta_0}{2}\right)+..\right.$$

Figure shows the relative errors using the power series. T_0 is the linear approximation, and $T_2...T_1$ include respectively the terms up to the 2nd... and up to the 10th powers.

For a swing of 180° the bob is balanced over its pivot point and so T = ∞ (keep in mind the pendulum is made of a rigid rod).

Potential energy and phase portrait of a simple pendulum. Note that the x-axis, being angle, wraps onto itself after every 2π radians.

For example, the period of a 1m pendulum at initial angle 10 degrees is $4\sqrt{\frac{1}{g}}F\left(\frac{\sin 10}{2},\frac{\pi}{2}\right)$ = 2.0102 seconds, whereas the approximation $2\pi\sqrt{\frac{1}{g}} = 2.0064$ that's about 1 second per swing (both examples use g = 9.80665 m/s²).

PHYSICAL INTERPRETATION OF THE IMAGINARY PERIOD

The Jacobian elliptic function that expresses the position of a pendulum as a function of time is a doubly periodic function with a real period and an imaginary period. The real period is of course the time it takes the pendulum to go through one full cycle. Paul Appell pointed out a physical interpretation of the imaginary period: if θ_0 is the maximum angle of one pendulum and 180° " θ_0 is the maximum angle of another, then the real period of each is the magnitude of the imaginary period of the other.

VECTOR PROPERTIES

It is the convention to print vectors in boldface type to distinguish them from scalars. Graphically speaking, an arrow can conveniently and effectively represent a vector. The length of the arrow is proportional to its magnitude while the tip of the arrow shows its direction.

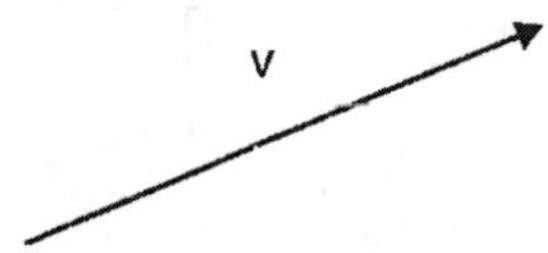

Figure

A vector can also be specified by its components:

$$v = \left[v_x, v_y, v_z \right]$$

The number of the components basically determine the dimension of a vector: 2-dimensional (2 components) vs. 3-dimensional (3 components). The components are related to the coordinate system in use. See the Coordinate Systems page for the details. Figure 2 shows vectors described in Cartesian coordinates.

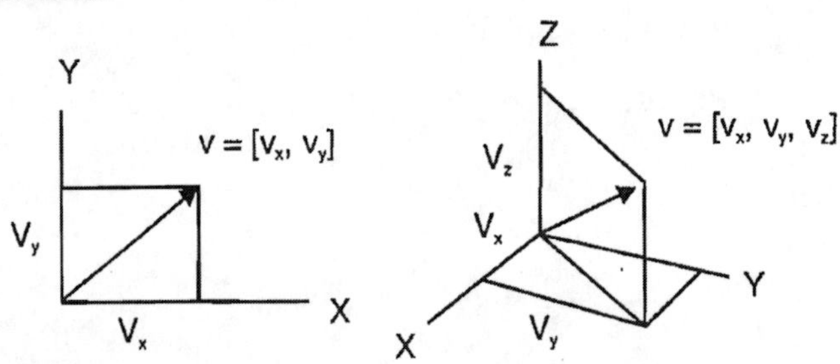

One can describe vector algebra in both graphically (using arrows) and with components.

The magnitude of the vector is the same to the length of the vector arrow shown in Figure:

$$v = |v| = \sqrt{v_x^2 + v_y^2 + v_z^2}$$

$$= \left(v_x^2 + v_y^2 + v_z^2\right)^{1/2}$$

A vector whose magnitude is unity (= 1) is called a unit vector. The unit vector of vector v shown in Figure can be obtained by dividing the vector with its magnitude.

$$\hat{v} = \frac{1}{v} v$$

The unit vector of the axes (unit coordinate vectors) can be expressed as

$$i = [1,0,0] j = [0,1,0] k = [0,0,1]$$

where i, j & k = the unit vectors of the X, Y & Z axes, respectively.

VECTOR ALGEBRA

For two vectors a & b, let

$$a \equiv [a_x, a_y, a_z]$$

$$b \equiv [b_x, b_y, b_z]$$

$$c \equiv [c_x, c_y, c_z]$$

If the two vectors are equal to each other, all the corresponding components are equal to each other as well:

$$a = b$$

$$[a_x, a_y, a_z] = [b_x, b_y, b_z]$$

$$a_x = b_x$$

$$a_y = b_y$$

$$a_z = b_z$$

Vectors of the same type can be added to yield the resultant vector. Adding two vectors is equivalent to adding the components:

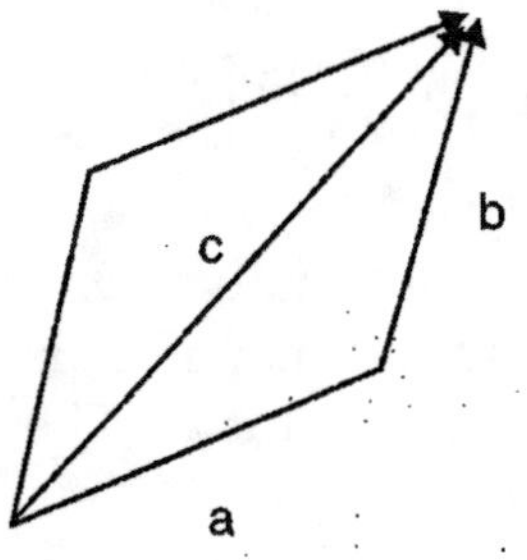

Figure

$$c = a + b$$
$$= \left[a_x, a_y, a_z\right] + \left[b_x, b_y, b_z\right]$$
$$= \left[a_x + b_x, a_y + b_y, a_z + b_z\right] = \left[c_x, c_y, c_z\right]$$

$$c_x = a_x + b_x$$
$$c_y = a_y + b_y$$
$$c_z = a_z + b_z$$

As shown in Figure, vectors a and b were connected tip-to-tale, forming two sides of a triangle. The third side is the resultant vector (c) drawn from the tale of the first vector to the tip of the last vector added. This is the so-called tip-to-tale (graphical) approach of vector addition. The sequence of addition is not important in this process (commutative):

$$a + b = \left[a_x, a_y, a_z\right] + \left[b_x, b_y, b_z\right]$$
$$= \left[a_x + b_x, a_y + b_y, a_z + b_z\right]$$
$$= \left[b_x + a_x, b_y + a_y, b_z + a_z\right] = b + a$$

Addition of more than two vectors can be also performed similarly, two at a time:

$$a + b + c = (a + b) + c = \left[a_x + b_x, a_y + b_y, a_z + b_z\right] + [c_x, c_y, c_z]$$
$$= \left[(a_x + b_x) + c_x, (a_y + b_y) + c_y, (a_z + b_z) + c_z\right]$$

$$= \left[a_x + b_x + c_x, a_y + b_y + c_y, a_z + b_z + c_z\right]$$

$$= \left[a_x + (b_x + c_x)', a_y + (b_y + c_y)', a_z + (b_z + c_z)\right]$$

$$= \left[a_x, a_y, a_z\right] + \left[b_x + c_x, b_z + c_y, b_z + c_z\right]$$

$$= \mathrm{a} + (\mathrm{b} + \mathrm{c})$$

Vector addition is associative. One can move the vectors freely to connect tip-to-tale as far as the magnitudes and the directions of the vectors are intact. The sequence in addition is not important.

When a vector is multiplied by a scalar, it magnifies:

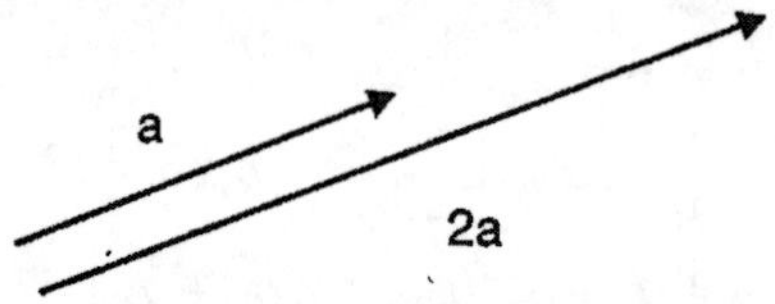

$$b = da$$

$$= d\left[a_x, a_y, a_z\right]$$

$$= \left[da_x, da_y, da_z\right] = \left[b_x, b_y, b_z\right]$$

$$b_x = da_x$$

$$\mathrm{b_y} = \mathrm{da_y}$$

$$\mathrm{b_z} = \mathrm{da_y}$$

where d = a scalar multiplier. The vector lengthens when $|d| > 1$ while it shortens if $|d| < 1$. If d is negative, the direction of the vector reverses. The operation shown above is commutative:

$$\mathrm{b} = \mathrm{da}$$

$$= d\left[a_x, a_y, a_z\right]$$

$$= \left[da_x, da_y, da_z\right] = \left[a_x d, a_y d, a_z d\right]$$

$$= \left[a_x, a_y, a_z\right] d$$

$$= \mathrm{ad}$$

Scalar multiplication is also distributive:

$$
\begin{aligned}
d(a + b) &= d\left[a_x + b_x, a_y + b_y, a_z + b_z\right] \\
&= \left[a\left(a_x + b_x\right), d\left(a_y + b_y\right), d\left(a_z + b_z\right)\right] \\
&= \left[da_x + db_x, da_y + db_y, da_z + db_z\right] \\
&= da + db
\end{aligned}
$$

In order to compute the displacement or to compute change in velocity, one needs to subtract one vector from another of the same kind. Vector subtraction is equivalent to subtraction of the corresponding components:

$$
\begin{aligned}
\mathrm{c} &= \mathrm{a} - \mathrm{b} \\
&= \mathrm{a} + (-1\mathrm{b}) \\
&= \left[a_x, a_y, a_z\right] + \left[-b_x, -b_y, -b_z\right] \\
&= \left[a_x - b_x, a_y - b_y, a_z - b_z\right] = \left[c_x, c_y, c_z\right]
\end{aligned}
$$

$$c_x = a_x - b_x$$
$$c_y = a_y - b_y$$
$$c_z = a_z - b_z$$

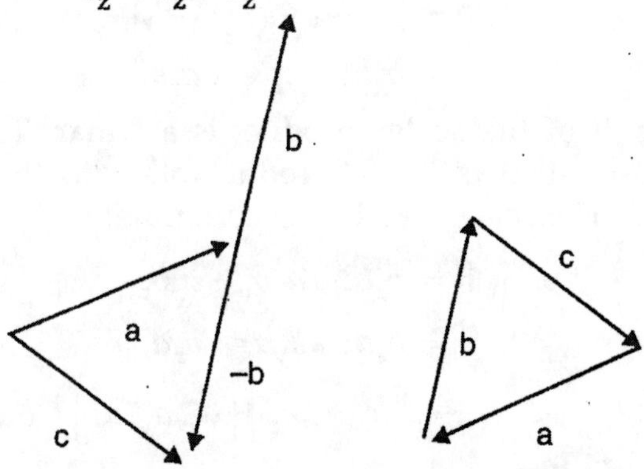

c = a − b　　　　c = a − b ⇒ b + c = a

Figure shows two equivalent graphical methods that find c. Vector subtraction is a special case of vector addition.

Now, let's revisit Figure. As shown in Figure, a vector can be expressed as the sum of the component vectors:

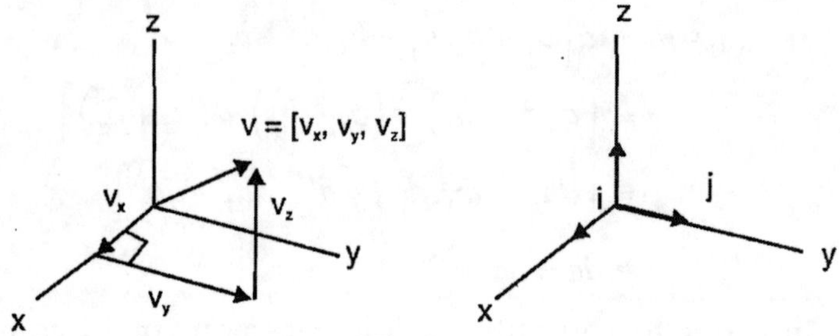

$$v_x = iv_x$$
$$v_y = jv_y$$
$$v_z = kv_x$$
$$v = [v_x, v_z, v_z]$$
$$= v_x + v_y + v_z = iv_x + jv_y + kv_z$$

where i, j & k = the unit coordinate vectors of the *X*, *Y* & *Z* axes, respectively.

VECTOR MULTIPLICATION

There are two different kinds of vector multiplication: the scalar product and the vector product. First, the scalar product is defined as:

$$a \cdot b = [a_x, a_y . a_z] \cdot [b_x, b_y, a_z]$$
$$= a_x b_x + a_y b_y + a_z b_z$$

The result of the scalar product is a scalar. The scalar product is also called as the dot product following the product symbol used. The scalar product is commutative:

$$a \cdot b = a_x b_x + a_y b_y + a_z b_z$$
$$= b_x a_x + b_y a_y + b_z a_z$$
$$= [b_x, b_y, b_z] \cdot [a_x, a_y, a_z] = b \cdot a$$

and distributive:

$$= [c_x, c_y, c_z] \bullet [a_x, a_y, a_z] + [c_x, c_y, c_z] \bullet [b_x, b_y, b_z]$$
$$= \mathbf{c} \bullet \mathbf{a} + \mathbf{c} \bullet \mathbf{b}$$

$$c \cdot (a + b) = [c_x, c_y, c_z] \cdot [a_x + b_x, a_y + b_y, a_z + b_z]$$

$$= c_x(a_x + b_x) + c_y(a_y + b_y) + c_z(a_z + b_z)$$

$$= c_x a_x + c_x b_x + c_y a_y + c_y b_y + c_z a_z + c_z b_z$$

$$= (c_x a_x + c_y a_y + c_z a_z) + (c_x b_x + c_y b_y + c_z b_z)$$

$$= [c_x, c_y, c_z] \bullet [a_x, a_y, a_z] +$$

$$[c_x, c_y, c_z] \square [b_x, b_y, b_z]$$

$$= c \bullet a + c \bullet b$$

From:

$$a \square a = a_x a_x + a_y a_y + a_z a_z$$

$$= a_x^2 + a_y^2 + a_z^2 = a^2$$

Square of the magnitude of a vector is the same to the scalar product of the vector with itself. Let's apply to the triangle formed by the three vectors shown in Figure:

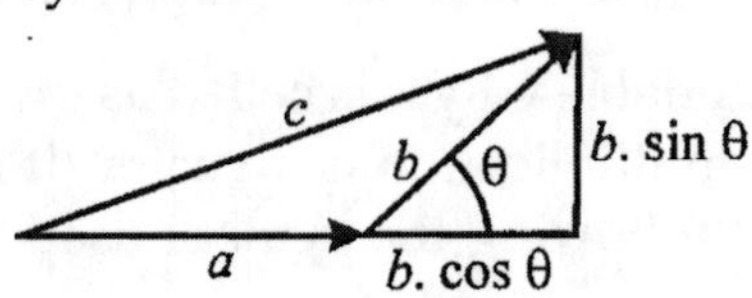

Figure 7

$$C^2 = c \bullet c(a+b) \bullet (a+b)$$

$$= a \bullet a + a \bullet b + b \bullet a + b \bullet b$$

$$= a \bullet a + 2a \bullet b + b \bullet b$$

$$= a^2 + 2a \bullet b + b^2$$

From the **Pythagorean Theorem:**

$$C^2 = (a + b\cos\theta)^2 + (b\sin\theta)^2$$

$$= a^2 + 2ab\cos\theta + b^2\cos^2\theta + b^2\sin^2\theta$$

$$= a^2 + 2ab\cos\theta + b^2(\cos^2\theta + \sin^2\theta)$$

$$= a^2 + 2ab\cos\theta + b^2$$

From Equation

$$a \bullet b = ab\cos\theta$$

where θ = the angle between the two vectors. Alternative definition of the scalar product. From:

$$i \bullet j = j \bullet k = k \bullet i = \cos 90° = 0$$
$$i \bullet i = j \bullet j = k \bullet k = \cos 0° = 1$$
$$v \bullet i = (v_x i + v_y j + v_z k) \bullet i = v_x$$
$$v \bullet j = (v_x i + v_y j + v_z k) \bullet j = v_y$$
$$v \bullet k = (v_x i + v_y j + v_z k) \bullet k = v_z$$

In other words, the components of a vector are its projections to the axes.

Another important and very useful vector multiplication is the vector product. The vector product is defined as:

$$a \times b = [a_x, a_y, a_z] \times [b_x, b_y, b_z]$$
$$= [a_y b_y - a_z b_y, a_z b_x - a_x b_z, a_x b_y - a_y b_x]$$

It is understandable why it is called as the vector product: the result of this multiplication is a vector. It is also called as the cross product following the symbol used.:

$$b \times a = [b_x, b_y, b_z] \times [a_x, a_y, a_z]$$
$$= [b_y a_z - b_z a_y, b_z a_x - b_x a_z, b_x a_y - b_y a_x]$$
$$= a \times b$$

and

$$c \times (a + b) = [C_x, C_y, C_z] \times [a_x + b_x, a_y + b_y, a_z + b_z]$$
$$= [C_y(a_z + b_z)\ C_z\ (a_y + b_y), C_z(a_x + b_x) - C_x(a_z + b_z),$$
$$C_x(a_y + b_y) - C_y(a_x + b_x)]$$
$$= c \times a + c \times b$$

From:

$$a \times a = [a_x, a_y, a_z] \times [a_x, a_y, a_z]$$
$$= [a_y a_z - a_z a_y, a_z a_x - a_x a_z, a_x a_y - a_y a_x]$$
$$= [0, 0, 0] = 0$$

In other words, the cross product of a vector with itself is always 0.

From:

$$|a \times b|^2 = (a \times b) \bullet (a \times b)$$
$$= (a_y b_z - a_z b_y)^2 + (a_z b_x - a_x b_z)^2 + (a_x b_y - b_y b_x)^2$$
$$= a_y^2 b_z^2 + a_z^2 b_y^2 + a_z^2 b_x^2 + a_x^2 b_z^2 + a_x^2 b_y^2 + a_y^2 b_x^2$$
$$- 2a_y a_z b_y b_z - 2a_x a_z b_x b_z - 2a_x a_y b_x b_y$$
$$= a_x^2 b_x^2 + a_x^2 b_y^2 + a_x^2 b_z^2 + a_y^2 b_x^2 +$$
$$a_y^2 b_y^2 + a_y^2 b_z^2 + a_z^2 b_x^2 + a_z^2 b_y^2 + a_z^2 b_z^2$$
$$-(a_x^2 b_x^2 + a_y^2 b_y^2 + a_z^2 b_z^2 + 2a_y a_z b_y b_z + 2a_x a_z b_x b_z + 2a_x a_y b_x b_y)$$
$$= (a_x^2 + a_y^2 + a_z^2)(b_x^2 + b_y^2 + b_z^2) - (a_x b_x + a_y b_y + a_z b_z)^2$$
$$= a^2 b^2 - (a \bullet b)^2$$
$$= a^2 b^2 - a^2 b^2 \cos^2 \theta = a^2 b^2 (1 - \cos^2 \theta)$$
$$= a^2 b^2 \sin^2 \theta$$
$$|a \times b| = ab \sin \theta$$

One thing very interesting here is the fact that the resulting vector from the cross product is perpendicular to the two vectors involved in the cross product:

$$(a \times b) \perp a$$
$$(a \times b) \perp b$$

since

$$(a \times b) \bullet a = [a_y b_z - a_z b_y, a_z b_x - a_x b_z, a_x b_y - a_y b_x] \bullet [a_x, a_y, a_z]$$
$$= (a_y b_z - a_z b_y) a_x + (a_z b_x - a_x b_z) a_y + (a_x b_y - a_y b_x) a_z$$
$$= 0$$
$$(a \times b) \bullet b = 0$$

The actual direction of the resulting vector follows the right-hand rule: the direction of the resulting vector is the direction a right-handed screw advances when it is rotated from the direction of the first vector to the direction of the second vector through the smallest angle between them.

From the right-hand rule:

$i \times j = kj \times i = -k$

$j \times k = ik \times j = -i$

$k \times i = ji \times k = -j$

$i \times i = j \times j = k \times k = 0$

DIFFERENTIATION

Let vector v be a function of time (t):

$$V(t) = [V_x(t), V_y(t), V_z(t)]$$
$$= V_x(t)i + V_y(t)j + V_z(t)k$$

The derivation operation is distributive so that the time-derivative of the vector is sum of the time-derivatives of the component vectors:

$$\frac{d(f(t)+g(t)+h(t))}{dt} = \frac{d(f(t))}{dt} + \frac{d(g(t))}{dt}\frac{d(h(t))}{dt}$$

$$\dot{V} = \frac{dV(t)}{dt} = \frac{dV_x(t)}{dt}i + \frac{dV_y(t)}{dt}j + \frac{dV_z(t)}{dt}k$$

$$= \dot{V}_x i + \dot{V}_y j + \dot{V}_z k = [\dot{V}_x, \dot{V}_y, \dot{V}_z]$$

holds because the unit coordinate vectors are constant. If the unit coordinate vectors are not constant (such as those in a rotating reference frame), does not hold.

In general, the following rules also hold:

$$\frac{d}{dt}(f(t).g(t)) = \frac{d(f(t))}{dt}.g(t) + f(t).\frac{d(g(t))}{dt}$$

$$\frac{d(ca)}{dt} = \frac{dc}{dt}a + c\frac{da}{dt}$$

$$\frac{d(a \bullet b)}{dt} = \frac{da}{dt} \bullet b + a \bullet \frac{ab}{dt}$$

$$\frac{d(a \times b)}{dt} = \frac{da}{dt} \times b + a \times \frac{ab}{dt}$$

since

$$\frac{d(ca)}{dt} = \lim_{\Delta t \to 0} \frac{c(t+\Delta t).a(t+\Delta t) - c(t).a(t)}{\Delta t}$$

$$= \lim_{\Delta t \to 0} \frac{c(t+\Delta t).a(t+\Delta t) - c(t).a(t+\Delta t) + c(t).a(t+\Delta t) - c(t).a(t)}{\Delta t}$$

$$= \lim_{\Delta t \to 0} \left\{ \frac{c(t+\Delta t) - c(t)}{\Delta t} .a(t+\Delta t) \right\}$$

$$+ = \lim_{\Delta t \to 0} \left\{ c(t) \frac{a(t+\Delta t) - a(t)}{\Delta t} \right\}$$

$$= \lim_{\Delta t \to 0} \frac{c(t+\Delta t) - c(t)}{\Delta t} a(t) + c(t) .$$

$$\lim_{\Delta t \to 0} \frac{a(t+\Delta t) - a(t)}{\Delta t}$$

$$= \frac{dc}{dt} a + c \frac{da}{dt}$$

MATRIX

Properties

A matrix is composed of several rows and columns of numbers:

$$\underset{(m \times n)}{a} = \begin{bmatrix} a_{11} & a_{12} & . & a_{1(n-1)} & a_{1n} \\ a_{21} & a_{22} & . & a_{2(n-1)} & a_{2n} \\ . & . & a_y & . & . \\ a_{(m-1)1} & a_{(m-1)2} & . & a_{(m-1)(n-1)} & a_{(m-1)n} \\ a_{m1} & a_{m2} & . & a_{m(n-1)} & a_{mn} \end{bmatrix}$$

As shown in figure, a m x n (dimension) matrix has m rows and n columns. The elements of the matrix are denoted by a_{ij} where i = the row number, and j = the column number. In biomechanics and motion analysis, 3 x 3 and 3 x 1 matrices are the most commonly used types.

A matrix of $m = n$ is called a square matrix. In a diagonal matrix which is a square matrix, all elements except the diagonal ones ($i = j$) are zero:

$$\underset{(m\times n)}{a} = \begin{bmatrix} a_{11} & 0 & . & 0 & 0 \\ 0 & a_{22} & . & 0 & 0 \\ . & . & a_n & . & . \\ 0 & 0 & . & a_{(m-1)(m-1)} & 0 \\ 0 & 0 & . & 0 & a_{mn} \end{bmatrix}$$

Matrices of the same dimension can be added or subtracted, element by element:

$$\underset{(3\times3)}{a} \pm \underset{(3\times3)}{b} = \begin{bmatrix} a_{11} & a_{12} & a_{13} \\ a_{21} & a_{22} & a_{23} \\ a_{31} & a_{32} & a_{33} \end{bmatrix} \pm \begin{bmatrix} b_{11} & b_{12} & b_{13} \\ b_{21} & b_{22} & b_{23} \\ b_{31} & b_{32} & b_{33} \end{bmatrix}$$

$$= \begin{bmatrix} a_{11} \pm b_{11} & a_{12} \pm b_{12} & a_{13} \pm b_{13} \\ a_{21} \pm b_{21} & a_{22} \pm b_{22} & a_{23} \pm b_{23} \\ a_{31} \pm b_{31} & a_{32} \pm b_{32} & a_{33} \pm b_{33} \end{bmatrix}$$

Matrix addition is commutative.

When a matrix is multiplied by a scalar, all the elements of the matrix magnify by the scalar value:

$$Ca = c\begin{bmatrix} a_{11} & a_{12} & a_{13} \\ a_{21} & a_{22} & a_{23} \\ a_{31} & a_{32} & a_{33} \end{bmatrix} = \begin{bmatrix} ca_{11} & ca_{12} & ca_{13} \\ ca_{21} & ca_{22} & ca_{23} \\ ca_{31} & ca_{32} & ca_{33} \end{bmatrix}$$

$$= \begin{bmatrix} a_{11} & a_{12} & a_{13} \\ a_{21} & a_{22} & a_{23} \\ a_{31} & a_{32} & a_{33} \end{bmatrix} C = aC$$

This operation is commutative.

Transpose of a m x n matrix is a n x m matrix whose columns are identical to the corresponding rows of the original matrix:

$$\underset{(m\times n)}{a} = \begin{bmatrix} a_{11} & a_{12} & . & a_{1(n-1)} & a_{1n} \\ a_{21} & a_{22} & . & a_{2(n-1)} & a_{2n} \\ . & . & . & a_y & . \\ a_{(m-1)1} & a_{(m-1)2} & . & a_{(m-1)(n-1)} & a_{(m-1)n} \\ a_{m1} & a_{m2} & . & a_{m(n-1)} & a_{(m-1)n} \end{bmatrix}$$

$$\underset{(m\times n)}{a} = \begin{bmatrix} a_{11} & a_{21} & . & a_{(m-1)3} & a_{m1} \\ a_{12} & a_{22} & . & a_{(m-1)2} & a_{n2} \\ . & . & a_y & . & . \\ a_{1(n-1)} & a_{2(n-1)} & . & a_{(m-1)(n-1)} & a_{m(n-1)} \\ a_{1n} & a_{2n} & . & a_{(m-1)n} & a_{mn} \end{bmatrix}$$

That the superscripted t, $()^t$, is used as the symbol for the transpose. Transposing suffices the following:

$$(a^t)^t = \begin{bmatrix} & . & \\ . & a_y & . \\ & . & \end{bmatrix}^t = \begin{bmatrix} & . & \\ . & a_y & . \\ & . & \end{bmatrix} = a$$

$$(a+b)^t = \begin{bmatrix} & . & \\ . & a_y + b_y & . \\ & . & \end{bmatrix}^t = \begin{bmatrix} & . & \\ . & a_y + b_y & . \\ & . & \end{bmatrix} = a^t + b^t$$

$$(ca)^t = \begin{bmatrix} & . & \\ . & ca_y & . \\ & . & \end{bmatrix}^t = \begin{bmatrix} & . & \\ . & ca_y & . \\ & . & \end{bmatrix} = ca^t$$

MATRIX MULTIPLICATION

One of the most useful properties of the matrix is the matrix multiplication. Imagine a system of linear equations:

$$a_{11}X + a_{12}Y + a_{13}Z = C_1$$

$$a_{21}X + a_{22}Y + a_{23}Z = C_2$$
$$a_{31}X + a_{32}Y + a_{33}Z = C_3 \qquad [7]$$

where a's & c's = scalars, and x, y & z = the unknowns. [7] can be simplified as:

$$\begin{bmatrix} a_{11} & a_{12} & a_{13} \\ a_{21} & a_{22} & a_{23} \\ a_{31} & a_{32} & a_{33} \end{bmatrix} \begin{bmatrix} X \\ Y \\ X \end{bmatrix} = \begin{bmatrix} C_1 \\ C_2 \\ C_3 \end{bmatrix} \qquad [8]$$

[8] is expressed in matrix multiplication form. The general form of matrix multiplication is

$$\underset{(m\times p)}{c} = \underset{(m\times n)}{a} \cdot \underset{(n\times p)}{b}$$

$$c_q = \sum_k a_{RA} b_{uj} \qquad [9]$$

or

$$\begin{bmatrix} . & . & . \\ . & C_y & . \\ . & . & . \end{bmatrix} = \begin{bmatrix} . & . & . & . & . \\ a_{i1} & a_{i2} & . & a_{i(n-1)} & a_{in} \\ . & . & . & . & . \end{bmatrix} \begin{bmatrix} . & b_{1i} & . \\ . & b_{2i} & . \\ . & . & . \\ . & b_{(p-1)}i & . \\ . & b_{pi} & . \end{bmatrix}$$

$$\begin{bmatrix} . & . & . \\ . & a_{i1}b_{1i} + a_{i2}b_{2i} + .. + a_{i(n-1)}b_{(n-1)i} + a_{in}b_{ni} & . \\ . & . & . \end{bmatrix}$$

Note that the number of columns in the left vector (n in) must be the same to the number of rows in the right matrix. The dimension of the resulting vector is m (rows of the left matrix) x p (columns of the right matrix) as shown in.

The matrix multiplication is not commutative:

$$\begin{bmatrix} & . & \\ . & a_{ij} & . \\ & . & \end{bmatrix} \begin{bmatrix} & . & \\ . & b_{ij} & . \\ & . & \end{bmatrix} = \begin{bmatrix} & . & \\ . & \sum_k a_{ij} b_{kj} & . \\ & . & \end{bmatrix}$$

$$\begin{bmatrix} & \cdot & \\ \cdot & b_{ij} & \cdot \\ & \cdot & \end{bmatrix}\begin{bmatrix} & \cdot & \\ \cdot & a_{ij} & \cdot \\ & \cdot & \end{bmatrix}=\begin{bmatrix} & \cdot & \\ \cdot & \sum_k a_{ij}b_{kj} & \cdot \\ & \cdot & \end{bmatrix}=\begin{bmatrix} & \cdot & \\ \cdot & \sum_k a_{uj}b_{m} & \cdot \\ & \cdot & \end{bmatrix}$$

but is distributive:

$$c(a+b)=\begin{bmatrix} & \cdot & \\ \cdot & C_{ij} & \cdot \\ & \cdot & \end{bmatrix}\begin{bmatrix} & \cdot & \\ \cdot & a_{ij}+b_{ij} & \cdot \\ & \cdot & \end{bmatrix}$$

$$=\begin{bmatrix} & \cdot & \\ \cdot & \sum_k C_m + b_{uj} & \cdot \\ & \cdot & \end{bmatrix}=\begin{bmatrix} & \cdot & \\ \cdot & \sum_k C_m a_{uj} + C_m b_{uj} & \cdot \\ & \cdot & \end{bmatrix}$$

$$=\begin{bmatrix} & \cdot & \\ \cdot & \sum_k C_m + b_{uj} & \cdot \\ & \cdot & \end{bmatrix}=\begin{bmatrix} & \cdot & \\ \cdot & \sum_k C_m b_{uj} & \cdot \\ & \cdot & \end{bmatrix}=ca+cb$$

Similarly:

$$(a+b)c=$$

$$=\begin{bmatrix} & \cdot & \\ \cdot & a_m + b_m & \cdot \\ & \cdot & \end{bmatrix}=\begin{bmatrix} & \cdot & \\ \cdot & C_{uj} & \cdot \\ & \cdot & \end{bmatrix}=\begin{bmatrix} & \cdot & \\ \cdot & \sum_k (a_m + b_{iu})C_{uj} & \cdot \\ & \cdot & \end{bmatrix}$$

$$=\begin{bmatrix} & \cdot & \\ \cdot & \sum_k a_m + C_{uj} & \cdot \\ & \cdot & \end{bmatrix}+\begin{bmatrix} & \cdot & \\ \cdot & \sum_k b_m C_{uj} & \cdot \\ & \cdot & \end{bmatrix}=ac+bc$$

Matrix multiplication is also associative:

$$a(bc) = \begin{bmatrix} & \cdot & \\ \cdot & a_m & \cdot \\ & \cdot & \end{bmatrix} + \begin{bmatrix} & \cdot & \\ \cdot & \sum_k b_{mk} C_{uj} & \cdot \\ & \cdot & \end{bmatrix}$$

$$= \begin{bmatrix} & \cdot & \\ \cdot & \sum_k b_{im} \left(\sum_k b_{mk} C_{uj} \right) & \cdot \\ & \cdot & \end{bmatrix} = \begin{bmatrix} & \cdot & \\ \cdot & \sum_k \left(\sum_k a_{im} b_{mk} \right) C_{uj} & \cdot \\ & \cdot & \end{bmatrix}$$

$$= \begin{bmatrix} & \cdot & \\ \cdot & \sum_k a_{im} b_{mk} & \cdot \\ & \cdot & \end{bmatrix} = \begin{bmatrix} & \cdot & \\ \cdot & C_{kj} & \cdot \\ & \cdot & \end{bmatrix} = (ab)c$$

For any scalar d:

$$d(ab) = (da)b = a(ab)$$

An identity matrix (I) is a square matrix whose diagonal elements are all 1 while the off-diagonal elements are all 0:

$$I = \begin{bmatrix} 1 & 0 & . & 0 & 0 \\ 0 & 1 & . & 0 & 0 \\ . & . & 1 & . & . \\ 0 & 0 & . & 1 & 0 \\ 0 & 0 & . & 0 & 1 \end{bmatrix}$$

Then:

$$aI = \begin{bmatrix} & \cdot & \\ \cdot & \sum_k a_{iu} \delta_{uj} & \cdot \\ & \cdot & \end{bmatrix} = \begin{bmatrix} & \cdot & \\ \cdot & a_{ij} & \cdot \\ & \cdot & \end{bmatrix} = a$$

$$la = \begin{bmatrix} & \cdot & \\ \cdot & \sum_k \delta_{im} a_{kj} & \cdot \\ & \cdot & \end{bmatrix} = \begin{bmatrix} & \cdot & \\ \cdot & a_{ij} & \cdot \\ & \cdot & \end{bmatrix} = a$$

where *d* shown in is the Kronecker delta:

$$\delta_{ij} = \begin{cases} 1 & (i = j) \\ 0 & (i \neq j) \end{cases}$$

$$(ab)^t = \begin{bmatrix} & \cdot & \\ \cdot & \sum_k a_{im} b_{uj} & \cdot \\ & \cdot & \end{bmatrix} = \begin{bmatrix} & \cdot & \\ \cdot & \sum_k b_{ui} a_{jk} & \cdot \\ & \cdot & \end{bmatrix} = b^t a^t$$

DETERMINANT

The determinant of a 2 x 2 matrix is defined as:

$$a = \begin{bmatrix} a_{11} & a_{12} \\ a_{21} & a_{22} \end{bmatrix}$$

$$\det a = \begin{bmatrix} a_{11} & a_{12} \\ a_{21} & a_{22} \end{bmatrix} = a_{11}a_{22} - a_{12}a_{21}$$

The determinant of a 3 x 3 matrix can be reduced to a series of the determinants of 2 x 2 matrices:

$$\det\, a = \begin{bmatrix} a_{11} & a_{12} & a_{13} \\ a_{21} & a_{22} & a_{23} \\ a_{31} & a_{32} & a_{33} \end{bmatrix}$$

$$= a_{11} \begin{vmatrix} a_{22} & a_{23} \\ a_{32} & a_{33} \end{vmatrix} - a_{12} = \begin{vmatrix} a_{11} & a_{23} \\ a_{31} & a_{33} \end{vmatrix} + a_{13} \begin{vmatrix} a_{21} & a_{22} \\ a_{31} & a_{32} \end{vmatrix}$$

$$= a_{11}(a_{22}a_{33} - a_{23}a_{32}) - a_{12}(a_{21}a_{33} - a_{23}a_{31}) + a_{13}(a_{21}a_{32} - a_{22}a_{31})$$

To generalize for any square matrix, a new matrix needs to be defined:

$$a = \begin{bmatrix} & \cdot & \cdot & \cdot & \\ \cdot & a_{(i-1)(j-1)} & a_{(i-1)j} & a_{(i-1)(j+1)} & \cdot \\ & a_{i(j-1)} & a_{ij} & a_{i(j+1)} & \cdot \\ \cdot & a_{(i+1)(j-1)} & a_{(i+1)j} & a_{(i+1)(j+1)} & \cdot \\ & \cdot & \cdot & \cdot & \end{bmatrix}$$

$$a_{ij} = \begin{bmatrix} & \cdot & \cdot & \\ \cdot & a_{(i-1)(j-1)} & a_{(i-1)(j+1)} & \cdot \\ \cdot & a_{(i+1)(j-1)} & a_{(i+1)(j+1)} & \cdot \\ & \cdot & \cdot & \end{bmatrix}$$

In other words, matrix a_{ij} is matrix a less the i-th row and the j-th column. Then, it can be generalize to

$$\det(a) = \sum_j (-1)^{1+i} a_{1i} \det(a_{1j})$$

is called a cofactor expansion across the first row of a. In fact, the determinant of a can be cofactor-expanded across any row or column:

$$\det(a) = \sum_j (-1)^{1+i} a_{ij} \det(a_{ij})$$

$$= \sum_j (-1)^{1+i} a_{ij} \det(a_{ij}) \quad]$$

The following properties of the determinant hold:

$$\det(a') = \det(a)$$

$$\det(ab) = \det(a)\det(b)$$

INVERSE MATRIX

The inverse matrix of a square matrix suffices the following relationship:

$$aa^{-1} = a^{-1}a = 1$$

where a^{-1} = the inverse matrix of a, and I = the identity matrix. For a square matrix to be invertible, its determinant must not be 0. The inverse matrix of square matrix a can be expressed as

$$a^{-1} = \frac{1}{\det(a)} \begin{bmatrix} & \cdot & \\ \cdot & a_{ij} & \cdot \\ & \cdot & \end{bmatrix}$$

$$a_{ij} = (-1)^{i+j} \det(a_p)$$

The matrix of cofactors of a on the right side in is called the adjugate of a. Note in that element a_{ij} of the adjugate is associated with $\det(a_{ji})$ rather than $\det(a_{ij})$.

A system of linear equations such as that in [8] can be generalized as

$$ax = b$$

where a = the known square coefficient matrix, b = the known column matrix, and x = the unknown parameter matrix. If matrix a is invertible, the unknown parameters can be obtained as follows:

$$a^{-1}(ax) = a^{-1}b$$

$$(a^{-1}a)x = a^{-1}b$$

$$Ix = x = a^{-1}b$$

In motion analysis, matrix a is generally not square. If matrix a is not a square matrix, the system of linear equations can be solved as

$$ax = b$$

$$(a^t a)x = a^t b$$

$$(a^t a)^{-1} (a^t a)x = (a^t a)^{-1} (a^t b)$$

$$x = (a^t a)^{-1} (a^t b)$$

Equation is the so-called least square method. See Least Square Method for more details.

Weighted Least Square Method

The actual least-square algorithm to be incorporated in a DLT programmes is more complex than of the DLT Method page. Typical least-square algorithm involves weights to assure a more stable set of zeros. For this, add the weight matrix to the DLT Method page:

$$XL = Y$$

$$WXL = WY$$

$$(X^tWX)L = X^tWY$$

$$(X^tWX)^{-1}\,(X^tWX)L = (X^tWX)^{-1}\,(X^tWY)$$

$$L = (X^tWX)^{-1}\,(X^tWY) \quad ,$$

where

$$W = \begin{bmatrix} W_1 & 0 & . & 0 & 0 \\ 0 & W_2 & . & 0 & 0 \\ . & . & . & . & . \\ 0 & 0 & . & W_{n-1} & 0 \\ 0 & 0 & . & 0 & W_n \end{bmatrix} ,$$

n = number of the equations, and W = diagonal weight matrix.

Camera Calibration

In the camera calibration, matrices X, Y & L

X =

$$\begin{bmatrix} \frac{X_1}{R_1} & \frac{Y_1}{R_1} & \frac{Z_1}{R_1} & \frac{1}{R_1} & 0 & 0 & 0 & 0 & \frac{-U_1X_1}{R_1} & \frac{-U_1Y_1}{R_1} & \frac{-U_1Z_1}{R_1} \\ 0 & 0 & 0 & 0 & \frac{X_1}{R_1} & \frac{Y_1}{R_1} & \frac{Z_1}{R_1} & \frac{1}{R_1} & \frac{-V_1X_1}{R_1} & \frac{-V_1Y_1}{R_1} & \frac{-V_1Z_1}{R_1} \\ \vdots & \vdots & \vdots & \vdots & \vdots & \vdots & \vdots & \vdots & \vdots & \vdots & \vdots \\ \frac{X_n}{R_n} & \frac{X_n}{R_n} & \frac{X_n}{R_n} & \frac{1}{R_n} & 0 & 0 & 0 & 0 & \frac{-U_nX_n}{R_n} & \frac{-U_nY_n}{R_n} & \frac{-U_nZ_n}{R_n} \\ 0 & 0 & 0 & 0 & \frac{X_n}{R_n} & \frac{Y_n}{R_n} & \frac{Z_n}{R_n} & \frac{1}{R_n} & \frac{-V_nX_n}{R_n} & \frac{-V_nY_n}{R_n} & \frac{-V_nZ_n}{R_n} \end{bmatrix}$$

$$
\left.\begin{matrix}
\frac{\xi r_1^2 R_1}{R_1} & \frac{\xi r_1^4 R_1}{R_1} & \frac{\xi r_1^6 R_1}{R_1} & \frac{(r_1^2+2\xi_1^2)R_1}{R_1} & \frac{\xi\eta_1 R_1}{R_1} \\
\frac{\eta r_1^2 R_1}{R_1} & \frac{\eta_1 r_1^4 R_1}{R_1} & \frac{\eta_1 r_1^6 R_1}{R_1} & \frac{\eta_1\xi_1 R_1}{R_1} & \frac{(r_1^2+2\eta_1^2)R_1}{R_1} \\
\vdots & \vdots & \vdots & \vdots & \vdots \\
\frac{\xi_n r_n^2 R_n}{R_n} & \frac{\xi_n r_n^4 R_n}{R_n} & \frac{\xi_n r_n^6 R_n}{R_n} & \frac{(r_n^2+2\xi_n^2)R_n}{R_n} & \frac{\xi_n\eta_n R_n}{R_n} \\
\frac{\eta_n r_n^2 R_n}{R_n} & \frac{\eta_n r_n^4 R_n}{R_n} & \frac{\eta_n r_n^6 R_n}{R_n} & \frac{\eta_n\xi_n R_n}{R_n} & \frac{(r_n^2+2\eta_n^2)R_n}{R_n}
\end{matrix}\right]
$$

$$L = [L_1 \;\; L_2 \;\; . \;\; L_{15} \;\; L_{16}]^t$$

$$Y = \left[\frac{U_1}{R_1} \;\; \frac{V_1}{R_1} \;\; . \;\; \frac{U_n}{R_n} \;\; \frac{V_n}{R_n}\right]^t \quad ,$$

where = the image-plane coordinates, = the object- space coordinates, L_1 - L_{16} = the DLT and additional parameters, n = the number of control points, and

$$\xi = U_i - U_0$$

$$\eta_i = V_i - V_0$$

$$r_i^2 = \xi_i^2 + \eta_i^2$$

$$R_i = L_9 X_i + L_{10} Y_i + L_{11} Z_i + 1 \quad ,$$

and $[U_0, V_0]$ = the coordinates of the principal point.

From equations of the DLT Method page:

$$\varepsilon_u = \frac{1}{R}(-L_1 X - L_2 Y - L_3 Z - L_4 + L_9 UX + L_{11} UZ)$$

$$- \Delta U + \frac{1}{R} U$$

$$\varepsilon_v = \frac{1}{R}(-L_5 X - L_6 Y - L_7 Z - L_8 + L_9 VX + L_{10} VY$$

$$+ L_{11} VZ) - \Delta V + \frac{1}{R} V ,$$

where $[\varepsilon_u, \varepsilon_v]$ = the random errors involved in the equation, and $[\Delta U, \Delta V]$ = optical errors.

From:

$$\partial\varepsilon_u = \frac{1}{R}(-L_1\partial X - L_2\partial Y - L_3\partial Z$$

$$+L_9\partial UX + L_9U\partial X + L_{10}\partial UY + L_{10}U\partial Y$$

$$+ L_{11}\partial UZ + L_{11}U\partial Z) + \frac{1}{R}\partial U$$

$$= \frac{1}{R}(-L_1\partial X - L_2\partial Y - L_3\partial Z + L_9U\partial X$$

$$+L_{10}U\partial Y + L_{11}U\partial Z) + \frac{1}{R}(L_9X + L_{10}Y + L_{11}Z + 1)\partial U$$

$$= \left(\frac{UL_9 - L_1}{R}\right)\partial X + \left(\frac{UL_{10} - L_2}{R}\right)\partial Y + \left(\frac{UL_{11} - L_3}{R}\right)\partial Z + \partial U$$

It was assumed that $L_1 - L_{11}$ and R are constants and

$$\partial\Delta U = \partial\Delta V = 0 .$$

From:

$$\sigma^2(\varepsilon_u) = \left(\frac{UL_9 - L_1}{R}\right)^2 \sigma^2(X) + \left(\frac{UL_{10} - L_2}{R}\right)^2 \sigma^2(Y)$$

$$+\left(\frac{UL_{11} - L_3}{R}\right)^2 \sigma^2(Z) + \sigma^2(Z) + \sigma^2(U)$$

$$\sigma^2(\varepsilon_v) = \left(\frac{VL_9 - L_5}{R}\right)^2 \sigma^2(X) + \left(\frac{VL_{10} - L_6}{R}\right)^2 \sigma^2(Y)$$

$$+\left(\frac{VL_{11} - L_7}{R}\right)^2 \sigma^2(Z) + \sigma^2(V) ,$$

where $[\sigma^2(\varepsilon_u), \sigma^2(\varepsilon_v)]$ = the variances of the equations, $[\sigma^2(U), \sigma^2(V)]$ = the variances of the image-plane coordinates, and $[\sigma^2(X), \sigma^2(Y), \sigma^2(Z)]$ = the variances of the object-space coordinates. The variance of the image-plane coordinates can be obtained by repeating digitization of the control points and computing the variance of their image-plane coordinates. The variance of the object-space coordinates can be obtained by repeating measurement of the real-life coordinates of the control points and computing their variance. It was assumed in [8] that x, y, z, u, and v are mutually independent.

It is a common practice to use the reciprocal of the variance as the weight associated with the equation:

$$W = \begin{bmatrix} \frac{1}{\sigma^2(\varepsilon_{u_1})} & 0 & . & 0 & 0 \\ 0 & \frac{1}{\sigma^2(\varepsilon_{v_1})} & . & 0 & 0 \\ . & . & . & . & . \\ 0 & 0 & . & \frac{1}{\sigma^2(\varepsilon_{u_n})} & 0 \\ 0 & 0 & . & 0 & \frac{1}{\sigma^2(\varepsilon_{v_n})} \end{bmatrix}.$$

Matrix X and Y in both contain R that is a function of L_9 - L_{11}. For this reason, an iterative approach is typically employed in the camera calibration. Here are a sketch of the iterative approach commonly used in programming:

1. In the first iteration, it is impossible to compute R since L_9, L_{10}, and L_{11} are not available. The coordinates of the principal point are not available as well. Therefore, the standard DLT with 11 parameters must be used. Set the weight matrix to the identity matrix 1.
2. From the second iteration on, compute R using the L_9, L_{10}, and L_{11} obtained from the previous iteration. Compute $[u_o, v_o]$ based on the L's obtained from the previous iteration. The DLT Method page for details.

Form the normal equation and solve the system for L_1 to L_{16}.

3. Repeat this procedure until all L's converge sufficiently. In either case, one must set a reasonable tolerance level for the convergence check.
4. Compute the variance-covariance matrix for L_1 - L_{11}. Since the system has only 10 independent factors, the parameters are mutually dependant. (See the Modified DLT page for details.) In other words, covariances among the parameters exist. The variance-covariance matrix for the L's will be used later in the reconstruction to compute the weights for the equations.

The variance-covariance matrix of the parameters can be expressed as

$$\sigma^2(L) = MSE(X^t WX)^{-1} \quad ,$$

where

$$MSE = \frac{Y^t WY - L^t X^t WY}{2n-16} \quad .$$

MSE is the mean square error from the least-square estimation. Since a control point provides two equations, the DOF (degree of freedom) for the *MSE* computation is $2n$ - 16 as shown in equation. The variance-covariance matrix of the parameters is symmetric and square (11 x 11):

$$\sigma^2(L) = \begin{bmatrix} \sigma(L_1,L_1) & \sigma(L_1,L_2) & . & \sigma(L_1,L_{10}) & \sigma(L_1,L_{11}) \\ \sigma(L_2,L_1) & \sigma(L_2,L_2) & . & \sigma(L_2,L_{10}) & \sigma(L_2,L_{11}) \\ . & . & . & . & . \\ \sigma(L_{10},L_1) & \sigma(L_{10},L_2) & . & \sigma(L_{10},L_{10}) & \sigma(L_{10},L_{11}) \\ \sigma(L_{11},L_1) & \sigma(L_{11},L_2) & . & \sigma(L_{11},L_{10}) & \sigma(L_{11},L_{11}) \end{bmatrix} .$$

The diagonal terms are the variances while the off-diagonal terms are the covariances.

Reconstruction

In the reconstruction, matrices X, Y & L

$$X = \begin{bmatrix} \frac{u^{(1)}L_9^{(1)} - L_1^{(1)}}{R^{(1)}} & \frac{u^{(1)}L_{10}^{(1)} - L_2^{(1)}}{R^{(1)}} & \frac{u^{(1)}L_{11}^{(1)} - L_3^{(1)}}{R^{(1)}} \\ \frac{\omega^{(1)}L_9^{(1)} - L_5^{(1)}}{R^{(1)}} & \frac{\omega^{(1)}L_{10}^{(1)} - L_6^{(1)}}{R^{(1)}} & \frac{\omega^{(1)}L_{11}^{(1)} - L_7^{(1)}}{R^{(1)}} \\ \vdots & \vdots & \vdots \\ \frac{U^{(m)}L_9^{(m)} - L_5^{(m)}}{R^{(m)}} & \frac{U^{(m)}L_{10}^{(m)} - L_2^{(m)}}{R^{(m)}} & \frac{U^{(m)}L_{11}^{(m)} - L_3^{(m)}}{R^{(m)}} \\ \frac{\omega^{(m)}L_9^{(m)} - L_5^{(m)}}{R^{(m)}} & \frac{\omega^{(m)}L_{10}^{(m)} - L_6^{(m)}}{R^{(m)}} & \frac{U^{(m)}L_{11}^{(m)} - L_7^{(m)}}{R^{(m)}} \end{bmatrix}$$

L = [X Y Z]t

$$X = \begin{bmatrix} \frac{L_4^{(1)} - v^{(1)}}{R^{(1)}} & \frac{L_8^{(1)} - \omega^{(1)}}{R^{(1)}} & . & \frac{L_4^{(m)} - v^{(m)}}{R^{(m)}} & \frac{L_8^{(1)} - \omega^{(m)}}{R^{(m)}} \end{bmatrix}^t,$$

where m = the number of cameras, and

$$v^{(i)} = u^{(i)} - \Delta u^{(i)}$$

$$\omega^{(i)} = v^{(i)} - \Delta v^{(i)}$$

$$\Delta u^{(i)} = \xi^{(i)}\left[L_{12}^{(i)} r^{(i)^2} + L_{13}^{(i)} r^{(i)^4} + L_{14}^{(i)} r^{(i)^6}\right] +$$

$$L_{15}^{(i)}\left[r^{(i)^2} + 2\xi^{(i)^2}\right] + L_{16}^{(i)}\xi^{(i)}\eta^{(i)}$$

$$\Delta V^{(i)} = \eta^{(i)}\left[L_{12}^{(i)} r^{(i)^2} + L_{13}^{(i)} r^{(i)^4} + L_{14}^{(i)} r^{(i)^6}\right]$$

$$+ L_{12}^{(i)}\eta^{(i)}\xi^{(i)} + L_{16}^{(i)}\left[r^{(i)^2} + 2\eta^{(i)^2}\right]$$

$$\xi^{(i)} = U^{(i)} - U_0^{(i)}$$

$$\eta^{(i)} = V^{(i)} - V_0^{(i)}$$

$$r^{2(i)^2} = \xi^{(i)^2} + \eta^{(i)^2}$$

$$R^{(i)} = L_9^{(i)} X^{(i)} + L_{10}^{(i)} Y^{(i)} + L_{11}^{(i)} Z^{(i)} + 1 \quad .$$

Although the weight matrix is again in the form of, the actual method to compute the weights is different from equations

$$\partial\varepsilon_u = \frac{1}{R}(-\partial L_1 X - \partial L_2 Y - \partial L_3 Z - \partial L_4$$

$$+ \partial L_9 UX + L_9 \partial UX + \partial L_{10} \partial UY + L_{10} \partial UY$$

$$+ \ \partial L_{11} UZ + L_{11} \partial UZ) + \frac{1}{R} \partial U$$

$$= \frac{1}{R}(-\partial L_1 X - \partial L_2 Y - \partial L_3 Z - \partial L_4 + \partial L_9 UX$$

$$+ \partial L_{10} UY + \partial L_{11} UZ \ + \ \frac{1}{R}(L_9 X + L_{10} Y + L_{11} Z + 1)\, \partial U$$

$$= -\frac{X}{R}\partial L_1 - \frac{Y}{R}\partial L_2 - \frac{Z}{R}\partial L_3 - \frac{1}{R}\partial L_4 + \frac{UX}{R}\partial L_9$$

$$+ \frac{UY}{R}\partial L_{10} + \frac{UZ}{R}\partial L_{11} + \partial U$$

$$\partial\varepsilon_v - \frac{X}{R}\partial L_5 - \frac{Y}{R}\partial L_6 - \frac{Z}{R}\partial L_7 - \frac{1}{R}\partial L_8 + \frac{VX}{R}\partial L_9$$

$$+ \ \frac{VY}{R}\partial L_{10} + \frac{VZ}{R}\partial L_{11} + \partial V \ .$$

It was assumed that x, y, and z are constants and $\partial \Delta U = \partial \Delta V = 0$. Therefore, the variances of the equations are

$$\sigma^2(\varepsilon_u) = \sum_{j=1}^{11}\sum_{j=1}^{11} U_i U_j \sigma(L_i, L_j) + \sigma^2(U)$$

$$\sigma^2(\varepsilon_v) = \sum_{j=1}^{11}\sum_{j=1}^{11} V_i V_j \sigma(L_i, L_j) + \sigma^2(V) \quad ,$$

where $\sigma(L_i, L_j)$ = the variance/covariance between the parameters, and

$$U = [U_1\ U_2 ..\ U_{10}\ U_{11}]$$

$$= \begin{bmatrix} \frac{-X}{R} & \frac{-Y}{R} & \frac{-Z}{R} & \frac{-1}{R} & 0 & 0 & 0 & 0 & \frac{UX}{R} & \frac{UY}{R} & \frac{UZ}{R} \end{bmatrix}$$

$$V = [V_1\ V_2 ..\ V_{10}\ V_{11}]$$

$$= \begin{bmatrix} 0 & 0 & 0 & 0 & \frac{-X}{R} & \frac{-Y}{R} & \frac{-Z}{R} & \frac{-1}{R} & \frac{VX}{R} & \frac{VY}{R} & \frac{VZ}{R} \end{bmatrix} .$$

It was assumed in that u and v are independent from L_1 - L_{11}. L_1 - L_{11} are mutually dependent. The weight matrix for the reconstruction, therefore, becomes

$$W = \begin{bmatrix} \frac{1}{\sigma^2\left(\varepsilon_u^{(1)}\right)} & 0 & . & 0 & 0 \\ 0 & \frac{1}{\sigma^2\left(\varepsilon_v^{(1)}\right)} & . & 0 & 0 \\ . & . & . & . & . \\ 0 & 0 & . & \frac{1}{\sigma^2\left(\varepsilon_u^{(m)}\right)} & 0 \\ 0 & 0 & . & 0 & \frac{1}{\sigma^2\left(\varepsilon_v^{(m)}\right)} \end{bmatrix},$$

where m = the number of cameras.

Again, an iterative approach must be used in solving the system:

1. In the first iteration, use the identity matrix as the weight matrix. Solve the system for x, y & z.
2. From the second iteration on, compute the weight vectors. Use the x, y & z from the previous iteration in computing the weights and R.
3. Stop iteration when x, y & z sufficiently converge. Set a proper tolerance level for this.

3

Waves and Vibrations

THE ONE DIMENSIONAL WAVE EQUATION

The one dimensional wave equation (with no source term) is

$$\frac{1}{c^2}\frac{\partial^2\phi}{\partial t^2}=\frac{\partial^2\phi}{\partial x^2}$$

where c is the wave speed. (From purely dimensional arguments, c must have dimensions $m\ s^{-1}$ and hence must represent a speed.)

The general solution, known as d'Alembert's solution, is

$$\phi(x,t)=F(x-ct)+E(x+ct)$$

which can be easily verified, using the chain rule. It is, however, so easy to derive that it is worthwhile going through the derivation: Introduce new variables

$u = x - ct,\ v = x + ct$

Then the wave equation can be written

$$\frac{\partial^2\phi}{\partial u\partial v}=0$$

Now, this immediately integrates to

$$\frac{\partial\phi}{\partial u}=h(u)$$

and then

$$\phi(u,v)=\int h(u)du+E(v)=F(u)+E(v)$$

$$=F(x-ct)+E(x+ct)$$

This can also be written in the form

$$\phi(x,t)=\hat{F}(t-x/c)+\hat{E}(t+x/c)$$

which is slightly more convenient in what follows.

The important points are that:

- $F(x - ct)$ represents a wave of shape $F(x)$ (at time $t = 0$) moving in the direction of increasing x at constant speed c. That is, $F(x - ct)$ represents the same shape as $F(x)$, but with the origin moved to $x = ct$, ie the same shape as $F(x)$ translated by ct units in the positive x direction.
- $E(x + ct)$ represents a wave of shape $E(x)$ (at time $t = 0$) moving in the direction of decreasing x at constant speed c. That is, $E(x + ct)$ represents the same shape as $E(x)$, but with the origin translated to $x = - ct$.

The three dimensional problem

The three dimensional wave equation problem is

$$\frac{1}{c^2}\frac{\partial^2\phi}{\partial t^2}=\Delta\phi+f(x,t),$$

together with a radiation condition. Specifically, waves should travel outwards from points where $f(x,t)\neq 0$ towards infinity rather than travel in from infinity.

The associated Green's function, $G(x,t;y,\tau)$ satisfies

$$\frac{1}{c^2}\frac{\partial^2 G}{\partial t^2}-\Delta_x G=\delta(t-\tau)\delta(x-y),$$

together with the radiation condition that disturbances should radiate away from the point of disturbance, $x = y$, rather than towards it.

For the usual reasons,

$$\phi(x,t) = \int_{-\infty}^{\infty} \int_{-\infty}^{\infty} \int_{-\infty}^{\infty} \int_{-\infty}^{\infty} G(x,t;y,\tau) f(y,\tau) d^3y \; d\tau.$$

If we introduce $z = x - y$ and $T = t - \tau$

$$\frac{1}{c^2}\frac{\partial^2 G}{\partial T^2} - \Delta_z G = \delta(T)\delta(z)$$

Since $\delta(z) = \delta(r)/4\pi r^2$, where $\tau = |z|$ it follows that $G = G(r, T)$. Thus

$$\frac{1}{c^2}\frac{\partial^2 G}{\partial T^2} - \frac{1}{r}\frac{\partial^2}{\partial r^2}(rG) = \frac{1}{4\pi r^2}\delta(T)\delta(r),$$

For $r \neq 0$ we have $\delta(r) = 0$, hence after multiplying through by r

$$\left(\frac{1}{c^2}\frac{\partial^2}{\partial T^2} - \frac{\partial^2}{\partial r^2}\right)(rG) = 0$$

which is the one-dimensional wave equation for $rG(r, T)$. Using the D'Alembert solution for the one-dimensional wave equation we conclude that

$$G(r,T) = \frac{F(T - r/c)}{4\pi r} + \frac{E(T + r/c)}{4\pi r}$$

for some functions $F(T - r/c)$ and $E(T + r/c)$. (The factor of $1/4\pi$ is introduced purely for convenience in what follows.)

The radiation condition that disturbances should radiate away from $x = y$ rather than towards it becomes the condition that disturbances should move away from $r = 0$ ($r = |z| = |x - y|$) rather than towards $r = 0$. Since $F(T - r/c)$ represents a wave moving away from $r = 0$ and $E(T + r/c)$ represents a wave moving towards $r = 0$, it follows that we must take

$$G(r,T) = \frac{F(T - r/c)}{4\pi r},$$

This solution is singular as $r \to 0$, so we define a "generalised version" of it by

$$G(r,T) = \lim_{r \to 0}\left(H(r-e)\frac{F(T-r/c)}{4\pi r}\right)$$

where the point r=0 is excluded, and use the result from section 5.4 that

$$\Delta z\left[\lim_{\tau \to 0}\left(\frac{h(r-\tau)f(r)}{4\pi r}\right)\right] = \frac{f''(r)}{4\pi r} - f(0)\delta(z).$$

This shows that

$$\Delta z G = \frac{F''(T-r/c)}{4c^2 \pi r} - f(T)\delta(z)$$

and, differentiating G twice with respect to T we obtain

$$\frac{\partial^2 G}{\partial T^2} = \frac{F''(T-r/c)}{4\pi r}$$

so that

$$\frac{1}{c^2}\frac{\partial^2 G}{\partial T^2} - \Delta_z G = F(T)\delta(z) = \delta(T)\delta(z).$$

Thus

$$F(T) = \delta(T)$$

and therefore

$$G(r,T) = \frac{\delta(T-r/c)}{4\pi r}.$$

Recalling that $T = t - \tau, \tau = |z| = |x-y|$

$$G(x,t;y,\tau) = \frac{1}{4\pi|x-y|}\delta\left(t-\tau-\frac{1}{c}|x-y|\right).$$

Note that G=0 except where $t = \tau + \frac{1}{c}|x-y|$.

An important point:

The delta function in the Green's function,

$$G(x,t;y,\tau)=\frac{1}{4\pi|x-y|}\delta\left(t=\tau-\frac{1}{c}|x-y|\right).$$

is a *scalar* delta function not a vector delta function. That is, it is a one dimensional delta function rather than a three dimensional delta function $\delta(x)$.

Retarded Potentials

The solution of the problem

$$\frac{1}{c^2}\frac{\partial^2\phi}{\partial^2 t^2}-\Delta_3\phi=f(x,t)$$

is

$$\phi(x,t)=\int_{-\infty}^{\infty}\int_{-\infty}^{\infty}\int_{-\infty}^{\infty}\int_{-\infty}^{\infty} f(y,\tau)$$

$$\frac{1}{4\pi|x-y|}\delta\left(t-\tau-\frac{1}{c}|x-y|\right)d^3y\,d\tau$$

and the τ integral picks out the value of the integrand at $\tau=t-|x-y|/c$ hence

$$\phi(x,t)=\frac{1}{4\pi}\int_{-\infty}^{\infty}\int_{-\infty}^{\infty}\int_{-\infty}^{\infty}\frac{f(y,t-|x-y|/c)}{|x-y|}d^3y.$$

This is referred to as the *Retarded Potential Integral.* Essentially it represents the superposition (ie, the integral) of disturbances at points y. The disturbance which eminates from point y at time t reaches the point at x time $t+|x-y|/c$ (ie, at t plus the time it takes the wave to travel from y to x). The disturbance is attenuated with the distance between y and x. The following example makes this idea somewhat clearer.

EXAMPLE:

Find the field generated by a time harmonic point source at the origin which is switched on at time t=0.

In terms of the wave equation the problem is to solve

$$\frac{1}{c}\frac{\partial^2\phi}{\partial t^2}-\Delta\phi H=(t)\delta(x)e^{iwt}.$$

Using the retarded potential integral we have

$$\phi(x,t)=\frac{1}{4\pi}\int_{-\infty}^{\infty}\int_{-\infty}^{\infty}\int_{-\infty}^{\infty}\frac{f(y,t-|x-y|/c)}{|x-y|}d^3y.$$

where $f(y,t)=\delta(y)H(t)e^{iwt}$ so

$$\phi(x,t)=\frac{1}{4\pi}\int_{-\infty}^{\infty}\int_{-\infty}^{\infty}\int_{-\infty}^{\infty}\frac{\delta(y)}{|x-y|}(t-$$

$$|x-y|/c)\exp(iw(t-|x-y|/c))d^3y$$

and, as the $\delta(y)$ term simply selects the value of the integrand at $y=0$,

$$\phi(x,t)=\frac{H(t-|x|/c)}{4\pi|x|}\exp(iw(t-|x|/c)).$$

The term $H(t-|x|/c)$

means that the disturbance is confined to the expanding sphere $|x|\le ct$

whose radius expands at the wave speed c. The surface of the sphere $|x|=ct$

represents the wave front of the expanding sphere of disturbance. Once the wave front has passed a point, (so is inside the sphere, ie, $|x|<ct$) the disturbance is simply

$$\phi = \frac{1}{4\pi|x|}\exp(iw(t-|x|/c))$$

and the phase of the oscillation at x differs from the phase of the oscillation at the source of the disturbance, 0, by a factor of $|x|/c$. This phase difference is simply the time it takes the disturbance to propagate from 0 to x, travelling at speed c.

In other words, when the disturbance reaches the point $|x|$ it has the same phase that it had when it left the origin. Put another way, it will take a time $|x|/c$ before a disturbance at the origin will propagate out to an observer at position x. When the disturbance does reach the observer at x it will have the same phase that it had when it left the origin.

The factor of $1/(4\pi|x|)$ says that the amplitude of the disturbance decays inversely with the distance from the source of the disturbance (at least it does so inside the sphere of disturbance).

The retarded potential integral simply says that we can regard a continuous source term $f(y/\tau)$ as an integral of a lot of point source terms. The effect of the continuous source at position x time t, ie, $\phi(x,t)$, is simply the sum of all disturbances from point sources y, time τ divided by the distance between x and y (ie, the $1/|x-y|$ term) with the phase adjustment for the time it takes the disturbance at point y to reach the point x (ie, the $t|x-y|$ term).

This has the same form as the Green's function for the Helmholtz equation (but with the opposite sign). This is not surprising since, for $|x| < ct$, we have

$$\frac{1}{c^2}\frac{\partial^2\phi}{\partial t^2} = -\frac{\omega^2}{c^2}\phi = -k^2\phi$$

and hence

$$\frac{1}{c^2}\frac{\partial^2\phi}{\partial t^2} = \Delta\phi = H(t)\delta(x)e^{i\omega t}$$

reduces to

$$(\Delta + k^2)\phi = -H(t)\delta(x)e^{i\omega t}$$

and writing $\phi = H(t)e^{i\omega t}\psi$ this becomes the problem for (minus) the Helmholtz Green's function;

$$(\Delta + k^2)\psi = -\delta(x)$$

Waves in one space dimension

The one dimensional wave equation, with a source term, is

$$\frac{1}{c^2}\frac{\partial^2\phi}{\partial t^2} = \frac{\partial^2\phi}{\partial xt_1^2} + f(x_1, t)$$

We can think of this as a three dimensional equation

$$\frac{1}{c^2}\frac{\partial^2\phi}{\partial t^2} = \Delta\phi + f(x_1, t)$$

where, since the source term depends only on x_1 and t, ϕ depends only on x_1 and t. Thus the solution can be written as

$$\phi(x_2, t) = \int_{\infty}^{-\infty} \int_{y_2=-\infty}^{\infty} \int_{y_2=-\infty}^{\infty}$$

$$\int_{y_2=-\infty}^{\infty} G_3(x, t; y, \tau) f(y_1 \tau) dy_1 dy_2 dy_3 d\tau$$

where $G_3(x,\ t;\ y, \tau)$ is the three dimensional Green's function;

$$G_3(x, t; y, \tau) = \frac{1}{4\pi|x-y|}\delta(t - \tau - |x-y|/c).$$

Now, since f depends only on y_1 and τ we can write

$$\phi(x_1, t) = \int_{\infty}^{\infty} \left(\int_{y_2=-\infty}^{\infty} \int_{y_2=-\infty}^{\infty} G(x, t; y, \tau) dy_2 dy_3 \right)$$

$f(y_1,\tau)dy_1 d\tau.$

If we define $G_1(x_1,t,y_1,\tau)$ by

$$G_1(x_1,t;y_1,\tau)=\int_{y_2=-\infty}^{\infty}\int_{y_2=-\infty}^{\infty} G_3(x,t;y,\tau)dy_2dy_3$$

then we have

$$\phi(x_1,t)=\int_{-\infty}^{\infty}\int_{\infty}^{\infty} G_1(x,t;y_1,\tau)f(y_1,\tau)dy_1 d\tau,$$

so that $G_1(x,t;y_1,\tau)$ must be the Green's function for, that is

$$\frac{1}{c^2}\frac{\partial^2 G_1}{\partial t^2}-\frac{\partial^2 G_1}{\partial {x_1}^2}=\delta(x_1-y_1)\delta(t-\tau).$$

This observation is known as the *method of descent;* we descend from the solution of the three dimensional problem to the solution of the one (or two) dimensional problem. It is equally valid for the Poisson, Helmholtz and Diffusion equations, although it is rather pointless in the case of the Diffusion equation (where we use the one dimensional Green's function to find the three dimensional one).

In order to find $G_1(x_1,t_1;y_1,\tau)$ we have to find the integral

$$G_1(x_1,t;y_1,\tau)=\int_{y_2=-\infty}^{\infty}\int_{y_2=-\infty}^{\infty} G_3(x,t;y,\tau)dy_2dy_3$$

First we write $z = x - y$ and $T = t - \tau$ to that the problem becomes

$$G_1(z_1,T)=\int_{y_2=-\infty}^{\infty}\int_{y_2=-\infty}^{\infty}\frac{1}{4\pi|z|}\delta(T-|z|/c)\,dz_2dz_3,$$

where $z_1 = x_1 - y_1$ (and $z_2 = x_2 - y_2$, $z_3 = x_3 - y_3$, but since all of x_2, x_3, y_2 and y_3 are going to be integrated away, this is not particularly important). Now introduce cylindrical polar coordinates in which z_1 is the axial direction and

$$z_2 = r\cos\theta, z_3 = r\sin\theta$$

so that

$$z_1^2 + z_2^2 + z_3^2 = z_1^2 + r^2$$

and

$$dz_2 dz_3 = r d\theta dr, dz_1 = dz_1.$$

with $-\infty < z_1 < \infty, 0 \leq r < \infty$ and $0 \leq \theta < 2\pi$. Then

$$|z| = \sqrt{z_1^2 + z_2^2 + z_3^2} = \sqrt{z_1^2 + r^2}$$

and hence

$$G_1(z_1, T) = \int_{r=0}^{\infty}\int_{\theta=0}^{2\pi} \frac{1}{4\pi\sqrt{z_1^2 + r^2}}\,\delta\left(T - \frac{1}{c}\sqrt{z_1^2 + r^2}\right) r d\theta d\tau.$$

As the integrand does not depend on θ and as

$$\int_0^{\infty} d\theta = 2\pi$$

this becomes

$$G_1(z_1, T) = \frac{1}{2}\int_0^{\infty} \frac{1}{\sqrt{z_1^2 + r^2}}\,\delta\left(T - \frac{1}{c}\sqrt{z_1^2 + r^2}\right) r dr$$

Now note that

$$\frac{d}{dr}\sqrt{z_1^2 + r^2} = \frac{r}{\sqrt{z_1^2 + r^2}}$$

so that if we put

$$q = \frac{1}{c}\sqrt{z_1^2 + r^2}, dq = \frac{1}{c}\frac{d}{dr}\left(\sqrt{z_1^2 + r^2}\right) dr$$

and note that when $r = 0$

$$q = \frac{1}{c}\sqrt{z_1^2} = |z_1|/c$$

this becomes

$$G_1(z_1,T)=\frac{c}{2}\int_{|z_1|/c}^{\infty}\delta(T-q)dq.$$

If $|z_1|/c > T$ the integral is zero (since $T - q$ is always negative in the integral) and if $|z/|/c < T$ the integral is one (since we are integrating the delta function across a zero of its argument $T - |z_1|/c$. Thus

$$G_1(z_1,T)=\frac{c}{2}H(T-|z_1|/c)$$

and since $z_1=x_1 - y_1, T=t-\tau$ we find that

$$G_1(x_1,t;y_1,\tau)=\frac{c}{2}H(t-\tau-|x_1-y_1|/c).$$

An Alternative Derivation of the 1D Green's Function

Consider the Green's function for the one space dimension wave equation;

$$\frac{1}{c^2}\frac{\partial^2 G}{\partial t^2}-\frac{\partial^2 G}{\partial x^2}=\delta(x-y)\delta(t-\tau).$$

If we put $z = x - y$ and $T=t-\tau$ this becomes

$$\frac{1}{c^2}\frac{\partial^2 G}{\partial T^2}-\frac{\partial^2 G}{\partial z^2}=\delta(z)\delta(T)$$

We want the solution $G(z,T)$ which has outgoing wave behaviour, that is, the solution for which waves move outwards from $z = 0$ towards infinity (as opposed to waves that move inwards from $z=\pm\infty$ towards $z = 0$).

Ignoring the $\delta(z)\delta(T)$ term (which is only nonzero if $z = 0$ and $T = 0$) the d'Alembert solution can be written as

$$G(z,T)=\hat{F}(T-z/c)+\hat{E}(T+z/c)$$

where

$$\hat{F}(T-z/c)$$

represents a wave travelling away from $z = 0$ towards $z = \infty$ and

$$\hat{E}(T + z/c)$$

represents a wave travelling towards $z = 0$ from $z = -\infty$. If we want a wave that travels away from its source, at $z = 0$, then for $z > 0$ we need the wave moving away from $z = 0$ towards $z = \infty$, that is

$$\hat{F} = (T - z/c)$$

and for $z < 0$ we need the wave moving away from $z = 0$ towards $z = -\infty$, that is

$$\hat{E}(T + z/c).$$

Now observe that $|z| = z$ for $z > 0$, and the appropriate wave behaviour is (for $z > 0$)

$$\hat{F} = (T - z/c)$$

and for $z < 0$, $|z| = -z$ and the appropriate wave behaviour is (for $z < 0$)

$$\hat{E}(T + z/c) = \hat{E}(T - |z|c)$$

Thus, to get outgoing wave behaviour we want a solution of the form

$$F(T - |z|/c).$$

Now substitute $G(z, T) = F(T - |z|/c)$ into

$$\frac{1}{c^2}\frac{\partial^2 G}{\partial T^2} - \frac{\partial^2 G}{\partial z^2} = \delta(z)\delta(T).$$

Then

$$\frac{1}{c^2}\frac{\partial^2 G}{\partial t^2} = \frac{1}{c^2}F''(T - |z|/c)$$

and

$$\frac{\partial G}{\partial x} = -\frac{\text{sgn}(z)}{c}F'(T - |z|/c)$$

since (d/dz)|z| = sgn(z) (see Chapter 1). Thus

$$\frac{\partial^2 G}{\partial x^2} = -\frac{2\delta(z)}{c}F'(T-|z|/c) + \left(\frac{-\text{sgn}(z)}{c}\right)^2 F''(T-|z|/c)$$

since $(d/dz)\text{sgn}(z) = 2\delta(z)$ (again, see Chapter1). $\text{sgn}(z)^2 =$ 1 Since whatever z is and $\delta(z)f(z) = \delta(z)f(0)$ this becomes

$$\frac{\partial^2 G}{\partial x^2} = -\frac{2\delta(z)}{c}F'(T) + \frac{1}{c^2}F''(T-|z|/c).$$

Hence

$$\frac{1}{c^2}\frac{\partial^2 G}{\partial T^2} - \frac{\partial^2 G}{\partial x^2} = \frac{2\delta(z)}{c}F'(T) = \delta(z)\delta(T)$$

so that

$$F'(T) = \frac{c}{2}\delta(T)$$

and hence

$$F(T) = \frac{c}{2}H(T)$$

Thus

$$G(z,T) = F(T-|z|/c) = \frac{c}{2}H(T-|z|/c$$

so that

$$G(x,t;y,\tau) = \frac{c}{2}H((t-\tau)-|x-y|/c).$$

The Two Dimensional Green's Function

We can use the method of descent to deduce the two dimensional Greens function

$$G_2(x_1, x_2, t; y_1, y_2, \tau)$$

for the two dimensional wave equation

$$\frac{1}{c^2}\frac{\partial^2 \phi}{\partial t^2} - \left(\frac{\partial^2 \phi}{\partial x_1^{\,2}} + \frac{\partial^2 \phi}{\partial x_2^{\,2}}\right) = f(x_1, x_2, t).$$

We find that

$$G_2(x_1,x_2,t;y_1,y_2,\tau)=\int_{y_2=\infty}^{\infty} G_3(x,t;y,\tau)dy^3.$$

Evaluating this integral is in principle similar to the calculation given above, but more tedious. Suffice it say that the integral can be done and, writing $x = (x_1, x_2)$ and $y = (y_2, y_2)$, we have

$$G_2(x,t;y,\tau)=\frac{H(t-\tau-|x-y|)}{2\pi\sqrt{(t-\tau)^2-|x-y|^2/c^2}}.$$

(In fact, the method of descent was invented not to find the one dimensional Green's function but rather, as the *easiest* way of finding the two dimensional Green's function!)

A few problems:

Example:

Consider the one dimensional version of the the example in §8.2. That is, find the solution of the one dimensional wave equation subject to a time harmonic oscillator at the origin which switchs on at t=0.

The problem is to solve

$$\frac{1}{c^2}\frac{\partial^2\phi}{\partial t^2}-\frac{\partial^2\phi}{\partial x^2}=H(t)\delta(x)e^{i\omega t}.$$

In terms of the Green's function $G(x,t;y,\tau)$ the solution is

$$\phi(x,t)=\int_{-\infty}^{\infty}\int_{-\infty}^{\infty} G(x,t;y,\tau)H(\tau)\delta(y)e^{i\omega\tau}\,dyd\tau.$$

Since $H(\tau)=0$ for $\tau<0$ and

$$\int_{-\infty}^{\infty} G(x,t;y,\tau)\delta(y)dy=G(x,t;0,\tau)$$

we see that

$$\phi(x,t)=\int_{-\infty}^{\infty} G(x,t;0,\tau)e^{i\omega\tau}\,d\tau.$$

From

$$G(x,t;0,\tau)=\frac{c}{2}H(t-\tau-|x|/c,$$

hence

$$\phi(x,t)=\frac{c}{2}\int_0^{\infty} H(t-\tau-|x|/c)e^{i\omega r}d\tau.$$

Now, put

$$q=t-\tau-|x|/c, dq=-d\tau$$

so

$$\phi(x,t)=\frac{c}{2}\int_{-\infty}^{t-|x|/c} H(q)\exp(i\omega(t-|x|/c-q))dq$$

If $t - |x|/c < 0$ then the integral is zero, since q is always less than zero and hence $H(q) = 0$ over the whole range of integration. All this means is that if $|x| > ct$ then the disturbance which switches on at time $t = 0$ and travels at speed c has not had time to reach the point x which is a distance $|x|$ from the origin. Thus there is zero disturbance at x.

If $t - |x|/c > 0$ then

$$\int_{-\infty}^{t-|x|/c} H(q)\exp(i\omega(t-|x|/c-q))dq$$

$$=\int_0^{t-|x|/c} \exp(i\omega(t-|x|/c-q))dq$$

since $H(q) = 0$ for $q < 0$ and $H(q) = 1$ for $q > 0$. Note that

$$\int_0^{t-|x|/c} \exp(i\omega(t-|x|/c-q))dq = \exp(i\omega(t-|x|/c-q))$$

$$\int_0^{t-|x|/c} c^{-i\omega q}dq$$

and hence

$$\int_0^{t-|x|/c} \exp(i\omega(t-|x|/c-q))dq=\frac{i}{\omega}(1-e^{i\omega(t-|x|/c}).$$

This means that if $|x| < ct$ then the disturbance has had time to propagate at speed c from the origin to x and it gives the effect of the disturbance.

In total we find that

$$\phi(x,t)=\frac{ic}{2\omega}H(t-|x|/c)(1-e^{i\omega(t-|x|/c}).$$

Thus in the region $|x| > ct$ there is no disturbance (it has not had time to propagate from the origin) and in the region $|x| < ct$ there is a disturbance given by

$$\frac{ic}{2\omega}(1-e^{i\omega(t-|x|/c}).$$

Since this disturbance takes time $t = |x|/c$ to travel from the origin to x, it has the same phase when it reaches x as it had when it left the origin.

Unlike the three dimensional version of this problem, there is no attenuation of the amplitude of the wave as it moves away from the origin.

An initial-value problem:

In practice we are usually more concerned with Initial Value Problems (IVPs) for the form

$$\frac{1}{c^2}\frac{\partial^2\phi}{\partial t^2}-\frac{\partial^2\phi}{\partial x^2}=0, t>0, \phi(x,0)=f(x), \frac{\partial\phi}{\partial t}(x,0)=g(x).$$

We can turn this IVP into a problem soluble via Green's functions by the usual method. That is, put

$$\psi(x,t)=H(t)\phi(x,t)$$

so that $\psi=\phi$ for $t > 0$. Then, as usual, we have

$$\frac{\partial\psi}{\partial t}=\delta(t)\phi(x,0)+H(t)\frac{\partial\phi}{\partial t},$$

recall that $\delta(t)\phi(x,t)=\delta(t)\phi(x,0)$,

$$\frac{\partial^2\psi}{\partial t^2}=\dot{\delta}(t)\phi(x,0)+\delta(t)\frac{\partial\phi}{\partial t}(x,0)+H(t)\frac{\partial^2\phi}{\partial t^2},$$

again, recall that $\delta(t)\frac{\partial\phi}{\partial t}(x,t)=\delta(t)\frac{\partial\phi}{\partial t}(x,0)$, and

$$\frac{\partial^2 \psi}{\partial t^2} = \frac{\partial^2}{\partial x^2} H(t)\phi(x,t) = H(t)\frac{\partial^2 \phi}{\partial x^2}.$$

Thus

$$\frac{1}{c^2}\frac{\partial^2 \psi}{\partial t^2} - \frac{\partial^2 \psi}{\partial x^2} = \frac{1}{c^2}$$

$$\left(\dot{\delta}(t)f(x) + \delta(t)g(x) + H(t)\left(\frac{\partial^2 \phi}{\partial t^2} - c^2\frac{\partial^2 \phi}{\partial x^2}\right)\right)$$

and hence

$$\frac{\partial^2 \psi}{\partial t^2} - c^2\frac{\partial^2 \psi}{\partial x^2} = \frac{1}{c^2}\dot{\delta}(t)f(x) + \delta(t)g(x)$$

For $t > 0$, and we have

$$\phi = \psi = \frac{1}{c^2}\int_{-\infty}^{\infty}\int_{-\infty}^{\infty} G(x,t;y,\tau)$$

$$(\dot{\delta}(\tau)f(y) + \delta(\tau)g(y))dyd\tau.$$

where G is the one dimensional Greens function. The term

$$\int_{-\infty}^{\infty}\int_{-\infty}^{\infty} G(x,t;y,\tau)(\delta(\tau)g(y)dyd\tau$$

$$= \int_{-\infty}^{\infty} G(x,t;y,0)g(y)dy$$

and we deal with the other term

$$\int_{-\infty}^{\infty}\int_{-\infty}^{\infty} G(x,t;y,\tau)\,\dot{\delta}(\tau)f(y)dyd\tau,$$

by noting that

$$\int_{-\infty}^{\infty} G(x,t;y,\tau)\dot{\delta}(\tau)d\tau = -\int_{-\infty}^{\infty}\frac{\partial G}{\partial \tau}(x,t;y,\tau)\delta(\tau)d\tau,$$

that is, using the definition of the derivative $\dot{\delta}(\tau)$ of the delta function $\delta(\tau)$. This shows that

$$\int_{-\infty}^{\infty} G(x,t;y,\tau)\dot{\delta}(\tau)d\tau = -\frac{\partial G}{\partial \tau}(x,t;y,0)$$

and hence that

$$\phi(x,t) = \frac{1}{c^2}\int_{-\infty}^{\infty}\left[G(x,t;y,0)g(y) - \frac{\partial G}{\partial \tau}(x,t;y,0)f(y)\right]dy.$$

Now

$$G(x,t;y,0) = \frac{c}{2}H(t-|x-y|/c),$$

$$\frac{\partial G}{\partial \tau}(x,t;y,0) = -\frac{c}{2}\delta(t-|x-y|/c)$$

so

$$\phi(x,t) = \frac{1}{2c}\left[\int_{-\infty}^{\infty} H(t-|x-y|/c)g(y)dy + \right.$$

$$\left.\int_{-\infty}^{\infty} \delta(t-|x-y|/c)f(y)dy\right].$$

Finally note that

$$t - |x-y|/c < 0 \text{ if } y < x - ct,\ y > x + ct$$
$$t - |x-y|/c = 0 \text{ if } y = x - ct,\ y = x + ct$$
$$t - |x-y|/c > 0 \text{ if } x - ct < y < x + ct$$

so that

$$\int_{-\infty}^{\infty} H(t-|x-y|/c)g(y)dy = \int_{x-ct}^{x+ct} g(y)dy,$$

and

$$\int_{-\infty}^{\infty} \delta(t-|x-y|/c)f(y)dy = c(f(x+ct) + f(x-ct))$$

so that, in total

$$\phi(x,t) = \frac{1}{2}[f(x,ct) + f(x-ct)] + \int_{x-ct}^{x+ct} g(y)dy.$$

If we write

$$\hat{g}(x) = \frac{1}{c}\int^{z} g(y)dy$$

then

$$\frac{1}{2c}\int_{x-ct}^{z+ct} g(y)dy = \frac{1}{2}[\hat{g}(x+ct) - \hat{g}(x-ct)]$$

and hence

$$\phi(x,t) = \frac{1}{2c}[f(x+ct) + \hat{g}(x+ct)] + \frac{1}{2}[f(x-ct) - \hat{g}(x-ct)]$$

which is in the form

$$\phi(x,t) = F(x-ct) + E(x+ct)$$

as expected...

A Final Example:

Find the solution of

$$\frac{1}{c^2}\frac{\partial^2\phi}{\partial x^2} - \frac{\partial^2\phi}{\partial x^2} = 0, t > 0$$

with

$\phi(x,0) = \sin(\pi x)$ for $-1 < x < 1$, $\phi(x, 0) = 0$ otherwise

and

$$\frac{\partial\phi}{\partial x}(x,0) = 0.$$

From the above, the solution has the form

$$\phi(x,t) = \frac{1}{2}[f(x+ct) + f(x-ct)] + \frac{1}{2c}\int_{x-ct}^{x+ct} g(y)dy$$

where in this particular case

$$f(x) = H(1-|x|)\sin(\pi x),\ g(x) = 0.$$

Hence

$$\phi(x,t) = \frac{1}{2}[H(1-|x-ct|\sin(\pi(x-ct))$$

$$+H(1-|x+ct|\sin(\pi(x+ct))]$$

The solution is shown in the following figure:

Figure: Solution as a function of x and t for $-3 < x < 3, 0 < t < 2$

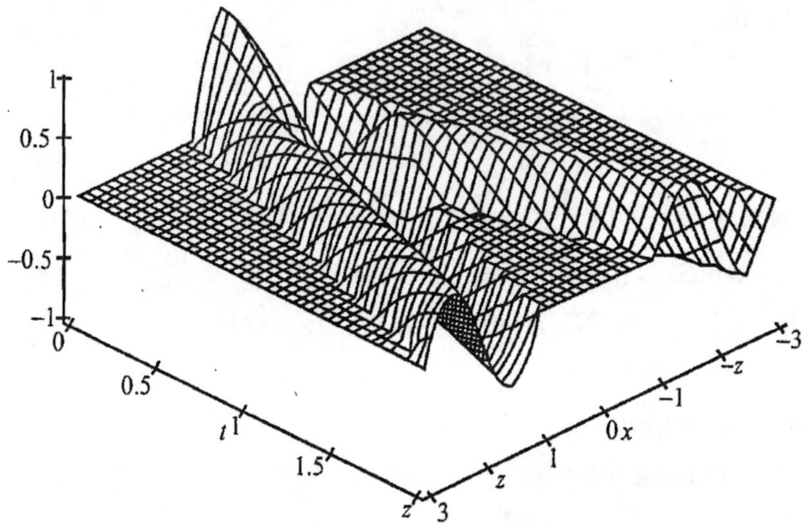

Exercises

1. Consider the wave equation problem

$$\frac{\partial^2 \phi}{\partial t^2} = \Delta\phi, \phi(x,0) = 0, \frac{\partial \phi}{\partial t}(x,0) = H(1-|x|),$$

where H(.) is the unit Heaviside function. Use the transformation $\psi(x,t) = H(t)\phi(x,t)$ to convert the problem into the form

$$\frac{\partial^2 \psi}{\partial t^2} - \Delta\psi = f(x,t).$$

Use the three dimensional Green's function to find the solution of this problem in integral form and hence show

$$\phi(0, t) = tH(1 - t), t > 0.$$

[*Note:* the delta function in the Green's function for the three-spatial dimensional wave equation is a ONE dimensional

Green's function, NOT a vector Green's function. You will have to use spherical polar co-ordinates to get the correct answer.]

2. Find the solution of the following problem

$$\frac{\partial^2\psi}{\partial t^2}=\frac{\partial^2\phi}{\partial x^2},\phi(x,0)=\cos(x),\frac{\partial\phi}{\partial t}(x,0)=2xe^{-x^2}$$

for $t > 0$. Show that this can be written as the sum of two waves, one moving to the left and one moving to the right, both at unit speed.

3. By writing $\psi = H(t)\phi$, show that the solution of

$$\frac{\partial^2\psi}{\partial t^2}-\frac{\partial^2\psi}{\partial x^2}=\delta(x)\sin(\omega t), t>0,$$

with

$$\phi(x,0)=\cos(x),\frac{\partial\phi}{\partial t}(x,0)=0$$

is

$$\phi(x,t)=\frac{1}{2}\left[\cos(x+t)+\cos(x-t)+\frac{1}{\omega}H(t-|x|)(1-\cos\right.$$

$(\omega(t\ |x|)))]$.

Eigenfunction

In mathematics, an eigenfunction of a linear operator, A, defined on some function space is any non-zero function f in that space that returns from the operator exactly as is, except for a multiplicative scaling factor. More precisely, one has

$$Af=\lambda f$$

for some scalar, λ, the corresponding eigenvalue. The solution of the differential eigenvalue problem also depends upon any boundary conditions required of f. In each case there are only certain eigenvalues $\lambda = \lambda_n$ (n = 1,2,3,...) that admit a corresponding solution for $f = f_n$ (with each f_n belonging to the eigenvalue λ_n) when combined with the boundary conditions.

The existence of eigenvectors is typically the most insightful way to analyse A.

For example, $f_k(x) = e^{kx}$ is an eigenfunction for the differential operator

$$A = \frac{d^2}{dx^2} - \frac{d}{dx}$$

for any value of k, with a corresponding eigenvalue $\lambda = k^2$ " k. If boundary conditions are applied to this system (e.g., $f = 0$ at two physical locations in space), then only certain values of $k = k_n$ satisfy the boundary conditions, generating corresponding discrete eigenvalues

$$\lambda_n = k_n^2 - k_n.$$

Applications

Eigenfunctions play an important role in many branches of physics. An important example is quantum mechanics, where the Schrödinger equation

$$i\eta\hbar \frac{\partial}{\partial t} \psi = H\psi$$

has solutions of the form

$$\psi(t) = \sum_k e^{-iE_k t/\hbar} \phi_k,$$

where ϕ_k are eigenfunctions of the operator H with eigenvalues E_k. The fact that only certain eigenvalues E_k with associated eigenfunctions ϕ_k satisfy Schrödinger's equation leads to a natural basis for quantum mechanics and the periodic table of the elements, with each E_k defining an allowable energy state of the system. The success of this equation in explaining the spectral characteristics of hydrogen is considered one of the great triumphs of 20th century physics.

Due to the nature of the Hamiltonian operator H, its eigenfunctions are orthogonal functions. This is not necessarily the case for eigenfunctions of other operators (such as the example A mentioned above). Orthogonal functions f_i, $i = 1, 2, \ldots$, have the property that

$$0 = \int f_i f_i$$

whenever $i \neq j$, in which case the set $\{f_i \mid i \in I\}$ is said to be linearly independent.

Eigenfunctions and Eigenvalues

An *eigenfunction* of an operator $\hat{A}$ is a function such that the application of $\hat{A}$ on f gives f again, times a constant.

$$\hat{A}f = kf$$

where k is a constant called the *eigenvalue*. It is easy to show that if $\hat{A}$ is a linear operator with an eigenfunction g, then any multiple of g is also an eigenfunction of $\hat{A}$.

When a system is in an *eigenstate* of observable A (i.e., when the wavefunction is an eigenfunction of the operator $\hat{A}$) then the expectation value of A is the eigenvalue of the wavefunction. Thus if

$$\hat{A}\psi(r) = a\psi(r)$$

then

$$\int \psi'(r)\hat{A}\psi(r)dr$$

$$\int \psi'(r)a\psi(r)dr$$

$$= a\int \psi'(r)\psi(r)dr$$

$$= a$$

assuming that the wavefunction is normalized to 1, as is generally the case. In the event that $\psi(r)$ is not or cannot be normalized (free particle, etc.) then we may use the formula

$$<A>=\frac{f\psi'(r)\hat{A}\psi(r)}{d}r\int\psi'(r)\psi(r)dr$$

What if the wavefunction is a combination of eigenstates? Let us assume that we have a wavefunction which is a linear combination of two eigenstates of $\hat{A}$ with eigenvalues a and b.

$$\psi=C_a\psi_n+c_a\psi_n$$

where $\hat{A}\psi_a=a\psi_a$ and $\hat{A}\psi_b=b\psi_b$. Then what is the expectation value of A?

$$<A>=\int\psi'\hat{A}\psi$$

$$\int[c_a\psi_a+c_b\psi_b]'\hat{A}[c_a\psi_a+c_b\psi_b]$$

$$=\int[c_a\psi_a+c_b\psi_b]'[ac_a\psi_a+bc_b\psi_b]$$

$$a|ca|^z\int\psi_a''\psi_a+bc_ac_b\int\psi_a''\psi_a+ac_b''c_a\int\psi_b''\psi_a+b|c_b|^2$$

$$\int\psi_b''\psi_b$$

$$a|c_a|^z+b|c_b|^z$$

assuming that ψ_a and ψ_b are orthonormal (shortly we will show that eigenvectors of Hermitian operators are orthogonal). Thus the average value of A is a weighted average of eigenvalues, with the weights being the squares of the coefficients of the eigenvectors in the overall wavefunction.

Hermitian Operators

As mentioned previously, the expectation value of an operator $\hat{A}$ is given by

$$<A>=\int\psi''(r)\hat{A}\psi(r)dr$$

and all physical observables are represented by such expectation values. Obviously, the value of a physical observable such as energy or density must be real, so we require $< A >$ to be real. This means that we must have

$$< A > = < A >^*,$$

or

$$\int \psi''(r)\hat{A}\psi(r)dr = \int (\hat{A}\psi(r))''\psi(r)dr$$

Operators which satisfy this condition are called *Hermitian*. One can also show that for a Hermitian operator,

$$\int \psi_1''(r)\hat{A}\psi_2(r)dr = \int (\hat{A}\psi_2(r))''\psi_2(r)dr$$

for any two states ψ_1 and ψ_2.

An important property of Hermitian operators is that their eigenvalues are real. We can see this as follows: if we have an eigenfunction of $\hat{A}$ with eigenvalue *u*, i.e. $\psi_a = a\psi_a$, then for a Hermitian operator

$$\int \psi_a''\hat{A}\psi_a = \int \psi_a(\hat{A}\psi_a)''$$

$$a\int \psi_a''\psi_a = a''\int \psi_a\psi_a''$$

$$(a - a'')\int |\psi_a|^3 = 0$$

Since $|\psi_a|^3$ is never negative, we must have either $a = a^*$ or $\psi_a = 0$. Since $\psi_a = 0$ is not an acceptable wavefunction, $a = a^*$, so *u* is real.

Another important property of Hermitian operators is that their eigenvectors are orthogonal (or can be chosen to be so). Suppose that ψ_a and ψ_b are eigenfunctions of $\hat{A}$ with eigenvalues *a* and *b*, with $a \neq b$. If $\hat{A}$ is Hermitian then

$$\int \psi_a^{*}\hat{A}\psi_b = \int \psi_b(\hat{A}\psi_a)^{*}$$

$$b\int \psi_a^{*}\psi_b = a^{*}\int \psi_b\psi_a^{*}$$

$$(b-a)\int \psi_a^{*}\psi_b = 0$$

since $a = a^*$ as shown above. Because we assumed $b \neq a$, we must have $\int \psi_a^{*}\psi_b = 0$, i.e. ψ_a and ψ_b are orthogonal. Thus we have shown that eigenfunctions of a Hermitian operator with different eigenvalues are orthogonal. In the case of degeneracy (more than one eigenfunction with the same eigenvalue), we can *choose* the eigenfunctions to be orthogonal. We can easily show this for the case of two eigenfunctions of $\hat{A}$ with the same eigenvalue. Suppose we have

$$\hat{A}\psi_j = j\psi_j$$

$$\hat{A}\psi_k = j\psi_k$$

We now want to take linear combinations of ψ_j and ψ_k to form two new eigenfunctions $\psi_{j'}$ and $\psi_{k'}$, where $\psi_{j'} = \psi_j$ and $\psi_{k'} = \psi_k + c\psi_j$. Now we want $\psi_{j'}$ and $\psi_{k'}$ to be orthogonal, so

$$\int \psi_{j'}^{*}\psi_{k'}$$

$$\int \psi_j^{*}(\psi_k + c\psi_j)$$

$$\int \psi_j^{*}\psi_k + c\int \psi_i^{*}\psi_j = 0$$

Thus we merely need to choose

$$C = -\frac{\int \psi_j^{*}\psi_k}{\int \psi_i^{*}\psi_j}$$

and we obtain orthogonal eigenfunctions. This Schmidt-orthogonalization procedure can be extended to the case of n-fold degeneracy, so we have shown that for a Hermitian operator, the eigenvectors can be made orthogonal.

OSCILLATOR

An oscillator is a mechanical or electronic device that works on the principles of oscillation: a periodic fluctuation between two things based on changes in energy. Computers, clocks, watches, radios, and metal detectors are among the many devices that use oscillators.

A clock pendulum is a simple type of mechanical oscillator. The most accurate timepiece in the world, the atomic clock, keeps time according to the oscillation within atoms. Electronic oscillators are used to generate signals in computers, wireless receivers and transmitters, and audio-frequency equipment, particularly music synthesizers. There are many types of electronic oscillators, but they all operate according to the same basic principle: an oscillator always employs a sensitive amplifier whose output is fed back to the input in phase. Thus, the signal regenerates and sustains itself. This is known as positive feedback. It is the same process that sometimes causes unwanted "howling" in public-address systems.

The frequency at which an oscillator works is usually determined by a quartz crystal. When a direct current is applied to such a crystal, it vibrates at a frequency that depends on its thickness, and on the manner in which it is cut from the original mineral rock. Some oscillators employ combinations of inductors, resistors, and/or capacitors to determine the frequency. However, the best stability (constancy of frequency) is obtained in oscillators that use quartz crystals.

In a computer, a specialized oscillator, called the clock, serves as a sort of pacemaker for the microprocessor. The clock frequency (or clock speed) is usually specified in megahertz (MHz), and is an important factor in determining the rate at which a computer can perform instructions.

OSCILLATOR

Oscillator is a processing library that generates series of waves. It also provides complex waves by mixing them each other.

This library is aimed for simulating rythm, pattern,Rÿepetition in the nature system. One of the strong feature of oscillator class is its recursive structure. You can mix waves. And you can also apply ocillators to other oscillator's parameter. (for example wave which is generated by oscillator controlls another oscillator's amplitude, period, offset).

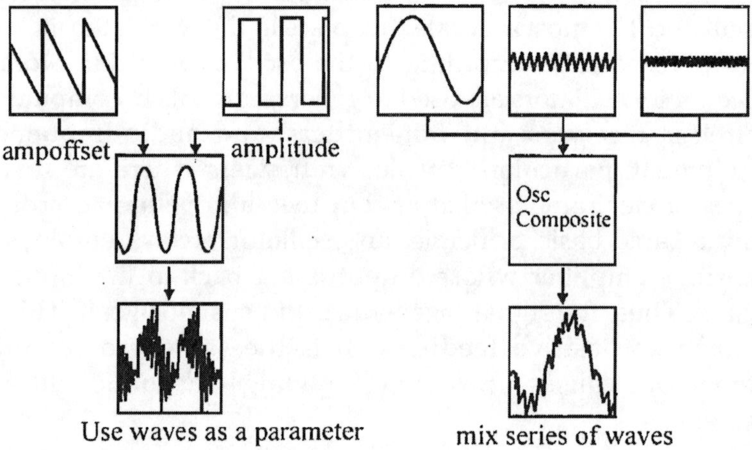

THE HARMONIC OSCILLATOR

The harmonic oscillator is a canonical system discussed in every freshman course of physics. Thus, you might skip this lecture if you are familiar with it.

Basic Equations of Motion and Solutions

The linearized equation of motion of an undamped and undriven pendulum is called a *harmonic oscillator*:

$$d^2\psi / dt^2 \omega 0^2 \psi = 0,$$

Nearly all oscillators and oscillations in physics are modeled by this equation of motion, at least in a first approximation, because it can be solved analytically. It is a linear second-order differential equation with constant coefficients. Such a differential equation can be solved easily by the ansatz $\psi = \exp(\lambda t)$, where λ is some constant which has to be calculated. Each d/dt puts a factor λ before the exponential function. Thus, a linear ordinary differential equation turns into a polynomial because the common exponential function (a nonzero factor) can be dropped. Thus, $\lambda + \omega 0^2 = 0$. The roots of this so-called characteristic polynomial are $\lambda = \pm i\omega 0$. Hence, we get *two* solutions. Because the equation of motion is *linear*, one gets new solutions by multiplying a known solution by an arbitrary factor or by adding a solution to another one. This so-called superposition principle of linear differential equations is very powerful. It says that *any* solution can be built up by only a few different types of "building blocks" called fundamental solutions. The solutions $\exp(-i\omega_0 t)$ and $\exp(+i\omega_0 t)$ are a set of two independent fundamental solutions of the harmonic oscillator. Thus, the general solution of reads

$$\psi(t) = C\exp(i\omega_0 t) + C^*\exp(-i\omega_0 t),$$

where C is a complex number and C^* denotes its conjugated complex. Because of the superposition principle, the set of fundamental solutions can be arbitrary, except that a fundamental solution has to be independent from the other ones. That is, each fundamental solution can not be written as a weighted sum of the other fundamental solutions. Instead of the complex valued exponential functions, one often chooses $\sin(\omega_0 t) = (\exp(i\omega_0 t) - \exp(-i\omega_0 t)/2i$ and $\cos(\omega_0 t) = (\exp(i\omega_0 t) + \exp(-i\omega_0 t)/2$ as another set of fundamental solutions. Therefore,

$$\psi(t) = A\sin(\omega_0 t) + B\cos(\omega_0 t),$$

is a general solution, too. The constants A and B can be expressed in term of C and vice versa.

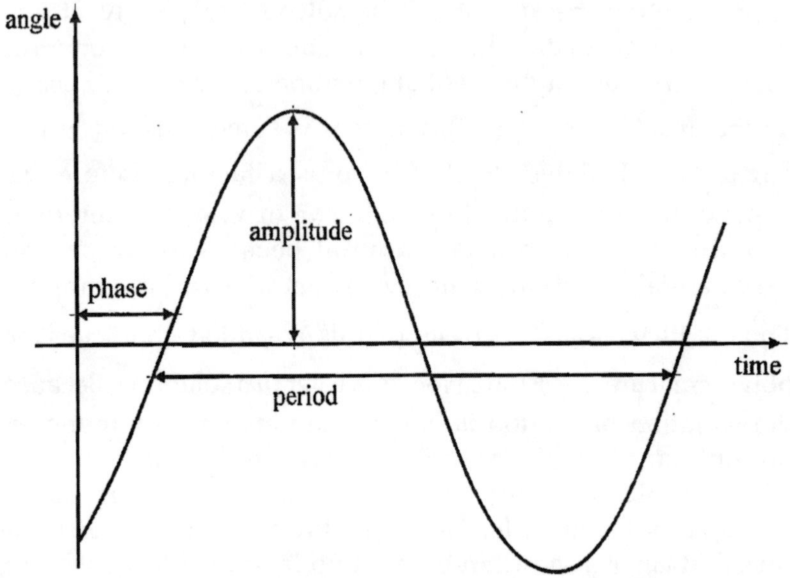

Using the porperties of trigonometric functions, the general solution can also be expressed in a physically more relevant way:

$$\psi(t) = \psi \max \sin(\omega_0 t - \psi),$$

where ψ_{max} and ψ are the amplitude and the phase of oscillation, respectively. Again they can be expressed in terms of the integration constants. Note, that these constants are determined by the initial conditions $\psi(0)$ and $d\psi/dt\,|_0$. The period of oscillation is $T_0 = 2\pi/\omega_0$. These three quantities are independent from each other. This isn't a triviality as nonlinear oscillations show. In fact, the independence of the frequency from the amplitude is a more subtile definition of a harmonic oscillator.

Damping

What happens if the harmonic oscilliator is damped? In this case, the equation of motions has an additional term which comes from the damping force:

$$d^2\psi/dt^2 + \gamma d\varphi/dt + \omega 0^2 \psi = 0.$$

Again, with the ansatz $\varphi = \exp(\lambda t)$, one gets a characteristic polynomial: $\lambda_2 + \gamma\lambda + \omega 0^2 = 0$. It has the solutions

$$\lambda_{\pm} = -\gamma/2 \pm (\gamma_2/4 - \omega 0^2)^{1/2}.$$

Depending on wether $\gamma_2/4$ is less than, equal, or greater than ω_{0^2}, one has to distinguish three cases:

The underdamped case: $\gamma/2 < \omega_0$

The eigenvalues $\lambda_{\pm}$ are a conjugated complex pair. The imaginary part $(\omega_0^2 - \gamma_{2/4})^{1/2}$ is just the frequency ω of a damped oscillation. The real part determines the rate of decay. The general solution reads

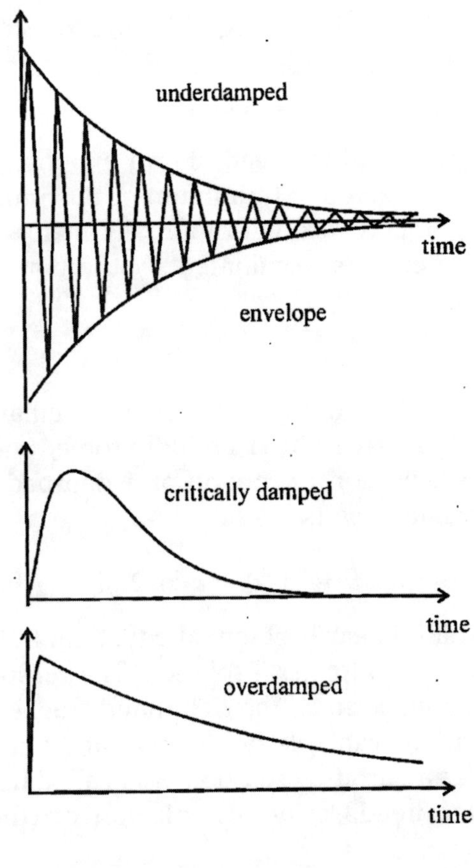

$$\varphi(t) = \varphi_{max} \exp(-\gamma t/2) \sin(\omega t + \psi),$$

The exponentially decaying amplitude of the oscillation is called the *envelope.*

The critically damped case: $\gamma/2 = \omega_0$

Both eigenvalues are identically. In this so-called *degenerated* case, the ansatz $\varphi = \exp(\lambda t)$ is not generally enough because it leads only to one solution. But for a fundamental system two independent solutions are needed. Where do we get a second solution? In such a case the ansatz $\varphi = t \exp(\lambda t)$ helps. Thus the general solution reads

$$\varphi(t) = (At + B) \exp(-\gamma t/2).$$

The overdamped case: $\gamma/2 > \omega_0$

Both eigenvalues are negative and real. The general solution reads

$$\varphi(t) = A + \exp\lambda(+ t) + A \exp\lambda(- t).$$

Any initial condition leads to an exponentially decay. There is only one extrema (minimum or maximum) contrary to the underdamped case with infinitely many maxima and minima. The decay is dominated by the eigenvalue most closest to zero.

Resonance

Oscillations in a damped harmonic oscillator decay to zero. When it is driven by a periodic force, one oscillation survives. The equation of motion of a damped and driven harmonic oscillator reads

$$d^2\psi/dt^2 + \gamma d\varphi/dt + \omega_0{}^2\psi = a \cos 2\pi ft.$$

This is still a linear differential equation, but the sum of two solutions are no longer a solution. The reason for that is the inhomogenous term on the right-hand side. Nevertheless, the superposition principle holds in a modified version. A solution of plus an arbitray solution (a so-called homogeneous solution). One needs only *one* solution of (the so-called particular solution) to generate any solution with the help of the homogeneous solution. Here, a particular solution can be found by the ansatz

$$\psi(t) = A\sin(2\pi ft) + B\cos(2\pi ft).$$

It turns into a sum with terms proportionally to $\cos(2\pi ft)$ and $\sin(2\pi ft)$. Because the equation should hold for any value of t, the sum of all cosinus terms and the sum of all sinus terms have to be zero independently. Hence, $[\omega_0^2 - (2\pi f)^2]\ B + 2\pi j\gamma A = a$ and $[\omega_0^2 - (2\pi f)^2]\ A - 2\pi j\gamma B = 0$, respectively. The solution of these equations reads

$$A = 2\pi j\gamma a/([\omega_{0^2} - (2\pi f)^2]^2 + (2\pi j\gamma)^2)$$

$$B = [\omega_{0^2} - (2\pi f)^2]a/([\omega_{0^2} - (2\pi f)^2]^2 + (2\pi f\gamma)^2).$$

In the long-time limit any solution of approaches the particular solution since any solution of decays to zero. The harmonic oscillator, therefore, oscillates *not* with its eigenfrequency ω_0 but with the frequency of the periodic force. The amplitude of oscillations depend on the driving frequency. It has its maximum when the driving frequency matches the eigenfrequency. This phenomenon is called resonance. To see this, rewrite equation in the form $\varphi(t) = \varphi_{max} \cos(2\pi ft - \psi)$, with

$$\varphi_{max} = a/([\omega_{0^2} - (2\pi f)^2]^2 + (2\pi j\gamma)^2)^{1/2}$$

and

$$\psi = \arctan(2\pi f\gamma/[\omega_{0^2} - (2\pi f)^2]).$$

In the underdamped case, the amplitude as a function of the driving frequency has a maximum near the eigenfrequency $f_0 = \omega_0/2\pi$ of the oscillator. The deviation is of quadratic order in γ. The maximum is $a/(\omega_0\gamma)$ in leading order of γ. This is by a factor ω_0/γ larger than the amplitude for $f \rightarrow 0$. That is, the amplitude of oscillation is larger than the amplitude of driving. This can be easily observed in the lab at the horizontally driven pendulum. The width of the so-called resonance line is proportional to γ. In the critically damped and overdamped case the resonance line disappears.

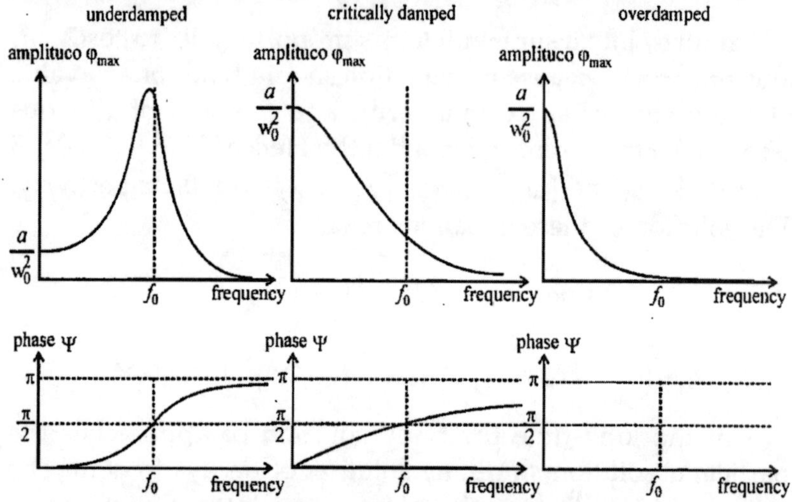

In the limit $f \to 0$ the phase of oscillation is identical with the driving phase. In resonance, the phase is just 90° behind the driving phase. At high frequencies the phase approaches 180°. Thus, the pendulum moves just opposite to the drive. Again, this can be easily observed in the lab at the horizontally driven pendulum.

Questions Worth to Think About

1. What happens with an undamped harmonic oscillator driven exactly by its eigenfrequency?
2. Another look at the dynamics of the damped and driven harmonic oscillator is the following one: Instead of discussing the solution as a function of time, discuss it as a function of the initial conditions. For example, calculate the angle and the angular velocity at a given time as a function of the angle and the angluar velocity for the previous period of driving. That is, given are the initial conditions $\varphi(0)$ and $d\varphi/dt|_0$. Calculate $\varphi(1/f)$ and $d\varphi/dt|_{1/f}$ as a

function of the initial conditions. This calculation leads to a Poincaré map or stroboscopic map.

NONLINEAR OSCILLATIONS

A pendulum is properly modelled by a harmonic oscillator only for small angles of elongation. Maybe you have observed in the lab that the period of oscillation increases with increasing amplitude of oscillation. Starting near the upside-down position, you will find that the period becomes much larger than for small-angle oscillations. In fact, the period approaches infinity in the limit $\varphi_{max} \to 180°$. In the virtual lab you will never reach this limit.

Even though the equation of motion of an undamped and undriven pendulum is *nonlinear*, one can calculate its frequency as a function of the amplitude φ_{max}. In order to solve this nonlinear differential equation

$$d^2\psi / dt^2 + \omega_{02} \sin\varphi = 0,$$

notice that the total energy

$$E = 1/2(d\varphi / dt)^2 - \omega_{02} \cos\varphi$$

is a constant during the motion of the pendulum. In the maximum elongation φ_{max}, the kinetic energy is zero. Thus,

$$E = -\omega_{02} \cos\varphi_{max}.$$

Solving for $d\varphi / dt$ leads to

$$d\varphi / dt = \pm[2(E + \omega_{02} \cos\varphi)]^{1/2}$$

This first-order differential equation can be solved by separating the independent variable t from the dependent one φ. That is, deal with the differential quotient as if it would be a real quotient $d\varphi$ divided by dt. Then, put all terms with φ on the left-hand side of the equation and all terms with t on the right-hand side:

$$d\varphi / [2(E + \omega_0^2 \cos\varphi)]^{1/2} = \pm dt.$$

To get the period of oscillations, integrate over a half cycle

$$\int_{-\varphi_{max}}^{\varphi_{max}} \frac{d\varphi}{\sqrt{2(E+\omega_0^2 \cos\varphi)}} = \int_0^{T/2} dt.$$

Because the integrand on the left-hand side is an even function in φ, you will get

$$T = \frac{2\sqrt{2}}{\omega_0} \int_{\varphi_0}^{\varphi_{max}} \frac{d\varphi}{\sqrt{\cos\varphi - \cos\varphi_{max}}}.$$

Note that E has been replaced by equation.

This integral can not be expressed by elementary functions like polynomials or trigonometric functions. This is possible only in the limit of $\varphi_{max} \to 0$, where the cosine function can be approximated by a Taylor series. Taking only the first and the second term, you will get an elementary integral leading to the well-known result of π/ω_0. For an arbitrary value of φ_{max}, the integral defines a so-called complete elliptic integral. It sounds like a naive trick to give an unsolvable integral a name. Indeed, it is a trick but a fruitful one because elliptic integrals appear quit often in mathematics and physics. Not as often as trigonometric functions of course, but often enough to give them a name, to find out their mathematical properties, to tabulate their values, and to include them into standard mathematical subroutine packages on computers. One distinguishes three different types of elliptic integrals. The second one is related to the circumference of an ellipse, that's why the name. In our case, the first one is relevant. The substitution $x = \sin(\varphi/2)/\sin(\varphi_{max}/2)$ turn into the canonical form of the complete elliptic integral of the first kind K

$$T = \frac{4}{\omega_0} \int_0^1 \frac{dx}{\sqrt{1-x^2}\sqrt{1-\sin^2(\varphi_{max}/2)x^2}} = \frac{4}{\varphi_0} K\left(\frac{\varphi_{max}}{2}\right).$$

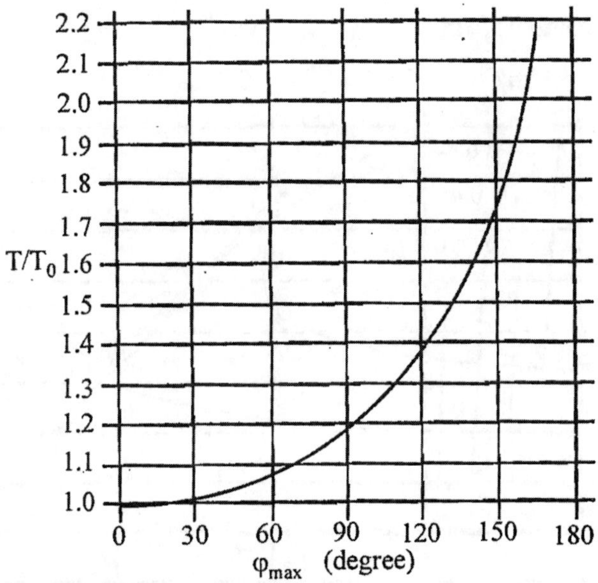

Here are a few properties of $K(\alpha)$:

$$K(0) = \pi/2,$$

$$K(\alpha \to \pi/2) \to \infty,$$

$$K(\alpha) = \pi/2(1 + \alpha_2/4 + ...).$$

Thus, in the limit $\varphi_{max} \to 0.$, T approaches to $T_0 = 2\pi/\omega_0$, i.e., the result of the harmonic oscillator. The plot shows the ratio T/T_0. Try to verify this curve by measurements! Go into the lab and measure the period of oscillation of the undamped and undriven pendulum with the stopwatch.

NONLINEAR RESONANCE

At the first glance, nonlinear resonance seems not to be so much different from the (linear) resonance of a harmonic oscillator. But both, the dependency of the eigenfrequency of a nonlinear oscillator on the amplitude and the nonharmoniticity of the oscillation lead to a behaviour that is impossible in harmonic oscillators, namely, the foldover effect and superharmonic resonance, respectively. Both effects are especially important in the case of weak damping.

The Foldover Effect

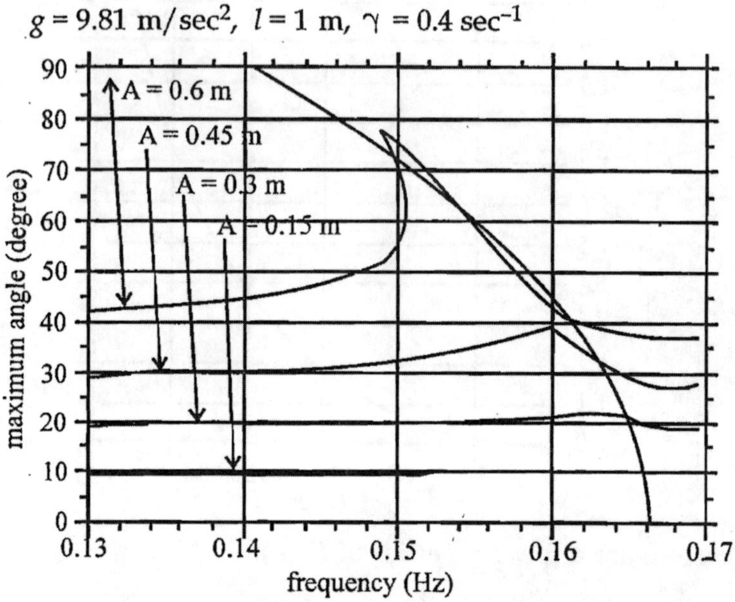

The foldover effect got its name from the bending of the resonance peak in a amplitude versus frequency plot. This bending is due to the frequency-amplitude relation which is typical for nonlinear oscillators. As an example, the figure on the left shows a sinusoidally driven pendulum. As long as the driving amplitude A is small, the resonance line is very well approximated by the result for an underdamped harmonic oscillator. For a stronger driving, the resonance lines "folds", leading to bistability and hysteresis. That is, the nonlinear oscillator oscillates either with a large amplitude or a small amplitude. Inbetween there is always an unstable periodic solution (shown as a dotted line). At the ends of the interval of bistablity this unstable limit cycle annihilates with one of its stable counterpart in a saddle-node bifurcation. The foldover effect also occurs in superharmonic resonance and parametric resonance.

The nonlinear resonance line can be approximated considerably well by the following heuristic approach: Take the resonance line of the harmonic oscillator and replace the eigenfrequency ω_0 by the frequency-amplitude relation $\omega_{res}(\varphi_{max}) = 2\pi/T$. This approximation assumes that the oscillation is still harmonic, i.e., sinusoidal, but the frequency depends on the amplitude. The figure shows that this approximation works quantitatively well even for large values of φ_{max}. By the way, the most easiest way to plot this approximation is to draw f as a function of φ_{max} because the equation for the resonance line is a quadratic in f^2.

For small values of the damping constant γ, the onset A_c of bistability is proportional to some power of γ. A power-law in the frequency-amplitude relation, i.e.,

$$(\omega_{res} - \omega_0) \sim (\varphi_{max})^s$$

leads to

$$A_C \sim \gamma 1 + 1/s,$$

because the width and the height of the resonance line are proportional to γ and A/γ, respectively.

Superharmonic Resonance

$g = 9.81$ m/sec^2, $l = 1$ m, $\gamma = 0.1$ sec^{-1}

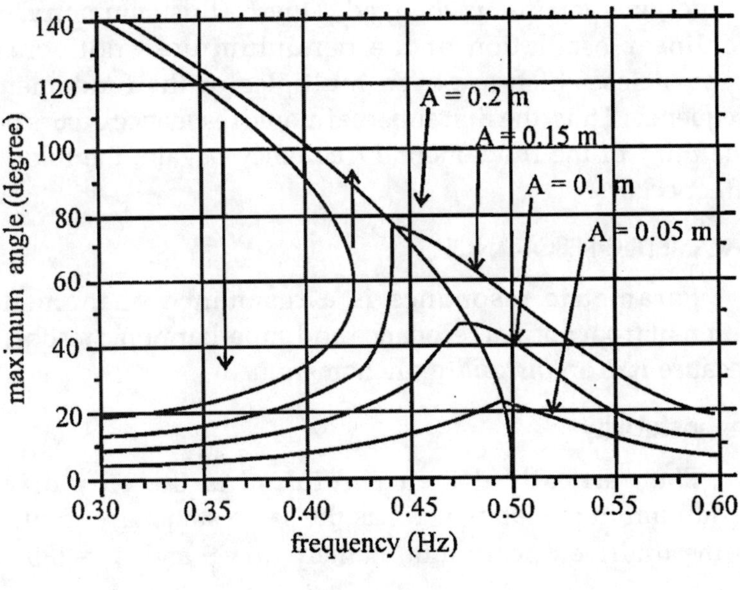

Nonlinear oscillators do not oscillate sinusoidal. Their oscillation is a sum of harmonic (i.e., sinusoidal) oscillations with frequencies which are integer multiples of the fundamental frequency (i.e., the inverse of the period of the nonlinear oscillation). This is the well-known theorem of Jean Baptiste Joseph Fourier (1768-1830) which says that periodic functions can be written as (infinite) sums (so-called Fourier series) of sinus and cosinus functions. Superharmonic resonance is simply the resonance with one of this higher harmonics of a nonlinear oscillation. In a plot of oscillation amplitude versus driving frequency, you can, therefore, expect additional resonance peaks. In general, they appear at driving frequencies which are integer fractions of the fundamental frequency.

If you try to find them for the pendulum driven by a periodic force, you have to choose relatively large driving amplitudes and low values for the damping constant. Nevertheless, you will not find superharmonic resonance at half the eigenfrequency of the pendulum! Why? The reason is *symmetry*. The undriven pendulum is *invariant* under reflection symmetry. That is, if you change the sign of φ, the equation of motion remains unchanged. Due to this symmetry, the nonlinear oscillation of the pendulum does not contain frequencies which are even multiples of the fundamental frequency. Thus, the first superharmonic resonance occurs near one third of the fundamental frequency. Again, the foldover effect occurs.

PARAMETRIC RESONANCE

Parametric resonance is a resonance phenomenon different from normal resonance and superharmonic resonance because it is an *instability* phenomenon.

The instability

The vertically driven pendulum is the only driven pendulum in the lab which has the same stationary solutions as the undriven pendulum, namely $\varphi = 0$ and $\varphi = 180°$. In the undriven case, these solutions are always stable and unstable, respectively. But vertical driving can change stability into instability and vice versa. On this page, mainly the destabilization of the normaly stable equilibrium of the pendulum (i.e., $\varphi = 0$) will be dicussed. Stabilization of an unstable upside-down pendulum will be mentioned only briefly.

In order to investigate the stability of a fixed point, you have to linearize the equation of motion around a fixed point. For $\varphi = 0$ one gets the (damped) Mathieu equation:

$$\frac{d^2\varphi}{dt^2} + \gamma\frac{d\varphi}{dt} + (\omega_0^2 + a\cos 2\pi f t)\varphi = 0$$

The same equation holds for $\varphi = 180°$ except that ω_0^2 has a minus sign. Thus, one studies the Mathieu equation for positive as well as negative values of ω_0^2.

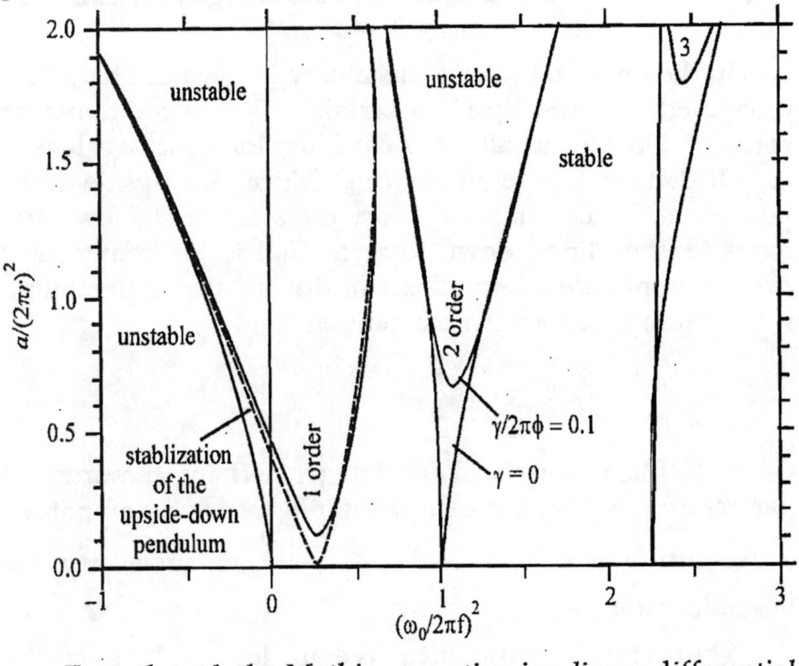

Even though the Mathieu equation is a linear differential equation, it can not be solved analytically in terms of standard functions. The reason is that one of the coefficients isn't constant but *time-dependent.* Fortunatly, the coefficient is periodic in time. This allows to apply the Floquet theorem. It says that in a linear differential equation or a system of linear differential equations there exists a set of *fundamental solutions* (from which one can build all other solutions) where all solutions can be written in the form

$$\varphi(t) = c(t)\exp(\lambda t), \text{ where } c(t + 1/f) = c(t).$$

The exponent is called Floquet exponent. It isn't uniquely defined because any factor $\exp(2\pi it/f)$ can be either absorbed in $c(t)$ or in $\exp(\lambda t)$. The solution is stable if the real part of all Floquet exponents are negative. If you are familiar with solid-state physics you will know the Floquet theorem as the *Bloch theorem.* In fact, the Mathieu equation is just the Schrödinger equation of an electron in a spatially periodic potential where the time variable is interpreted as a space variable. For a fixed value of a (i.e., the strength of the potential), the stable interval of ω_0^2 corresponds to the bands in the energy spectrum. The unstable intervals are just the band gaps.

In the figure the area of instability is shown in blue. It is calculated numerically for a certain value of the damping constant. Note the small area of stability for *negative* values of ω_0^2. It denotes the possibility of *stabilizing* the upside-down pendulum. In the undamped case the areas of stability form tongs (dashed lines) down to zero. That is, an infinitesimal driving amplitude a destabilizes the down-hanging pendulum if the parametric resonance condition

$$f_0 \equiv \frac{\omega_0}{2\pi} = \frac{f}{2}n$$

is fulfilled, where n is an integer defining the order of parametric resonance. In case of damping, a driving amplitude a exceeding a critical value $a_c \sim \gamma^{1/n}$ is necessary for destabilization.

There is a simple intuitive understanding of the parametric resonance condition. And maybe you already have such intuitive knowledge! Imagine you are on a fair and you want to swing a swing boat. Standing in the boat you have to go down with your body when reaching a maximum because you want to speed the boat up by putting an additional acceleration to the acceleration of gravity. If you do that near the forward and the backward maximum of oscillation, you are just realizing first-order parametric resonance because you are moving periodically up and down with a frequency just twice the frequency of the swing boat. If you go down only every *n*th maximum you will have parametric resonance of order *n*.

The onset of first-order parametric resonance can be approximated analytically very well by the ansatz:

$$\varphi = c + \exp(i\pi ft) + c - \exp(-i\pi ft)$$

This ansatz is a truncated Fourier series of the periodic function $c(t)$ of the Floquet theorem. It neither decays nor grows. Thus the amplitude a of driving has to be just the critical one a_c. Put this ansatz into and neglect the terms $\exp(\pm 3i\pi ft)$. Now, gather all terms having the factor $\exp(\pm i\pi ft)$ and you will get

$$\left[\omega_0^2 - (\pi f)^2 \pm i\gamma\pi f\right] c \pm + \frac{a_c}{2} c \mp = 0.$$

These two equations have a nontrivial solution for c_+ and c_- only if

$$a_c = 2\sqrt{\left[\omega_0^2 - (\pi f)^2 + (\gamma\pi f)^2\right]}$$

This instability threshold has a minimum just at the parametric resonance condition $f = \omega_0 / \pi$ (plus a correction of second order in γ). The minimum reads $a_c = 2\pi j \gamma$.

Parametrically Excited Oscillations

$g = 9.81\ \text{m/sec}^2,$

$l = 1\ \text{m}, \quad = 0.1\ \text{sec}^{-1}, \quad A = 0.07\ \text{m}$

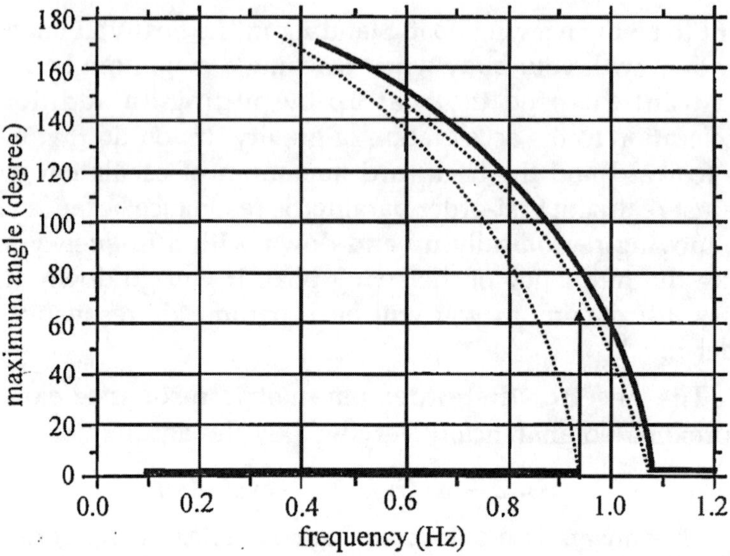

In parametric resonance the amplitude of the unstable solution grows exponentially to infinity. Damping does not help to saturate this growth contrary to normal resonance caused by an additive driving force. Thus, the *nonlinearities* in the parametrically driven pendulum are necessary for saturation. This saturation is caused by the fact that nonlinear oscillators have in general an amplitude-dependent eigenfrequency. The growth of the parametrically excited oscillation will shift the eigenfrequency out of resonance.

Thus a good approximation of the amplitude φ_{max} of the oscillation excited in first-order parametric resonance would be Equation. where ω_0 is replaced by $\varphi_{res}(\varphi_{max})$. The figure compares this approximation (green curves) with numerical solutions (red symbols). The approximation is indeed marvelous! For $\varphi_{max} \rightarrow 0$ the green curves end at the boundaries of the instability interval. In the language of nonlinear dynamics they mark supercritical and subcritical period doubling bifurcations. Thus one curve (dotted line) corresponds to an unstable solution. It is inbetween the stable down-hanging equilibrium and the stable oscillation. This bistability is responsible for hysteresis. Again as for nonlinear resonance a foldover effect has bend the resonance curves.

DESIGN AND CONSTRUCTION OF RADIO FREQUENCY OSCILLATORS

Some people regard the design of RF Oscillators to be something akin to a "black art" and after many years of swearing at "cranky" oscillators I'm not all too sure they are all that wrong. I suggest you ensure you remember this old saying:

"Amplifiers oscillate and oscillators amplify" - unknown

Principles of Oscillator operation

Every oscillator has at least one active device (smarties don't complicate matters for me - just read on) be it a transistor or even the old valve. This active device and, for this tutorial we'll stick to the humble transistor, acts as an amplifier. There is nothing flash about that. For this first part of the discussion we will confine ourselves to LC Oscillators and I'll keep the maths to an absolute minimum.

At turn on, when power is first applied, random noise is generated within our active device and then amplified. This noise is fed back positively through frequency selective circuits to the input where it is amplified again and so on, a bit like my childhood project. Ultimately a state of equilibrium is reached where the losses in the circuit are made good by consuming power from the power supply and the frequency of oscillation is determined by the external components, be they inductors and capacitors (L.C.) or a crystal. The amount of positive feedback to sustain oscillation is also determined by external components.

Hartley Oscillator

I decided to lead off with this one for the simple reason it's my favourite. Recently it was discussed that your favourite oscillator was likely the one which worked best for you and I think that is quite true. So here it is in it's most simplified form.

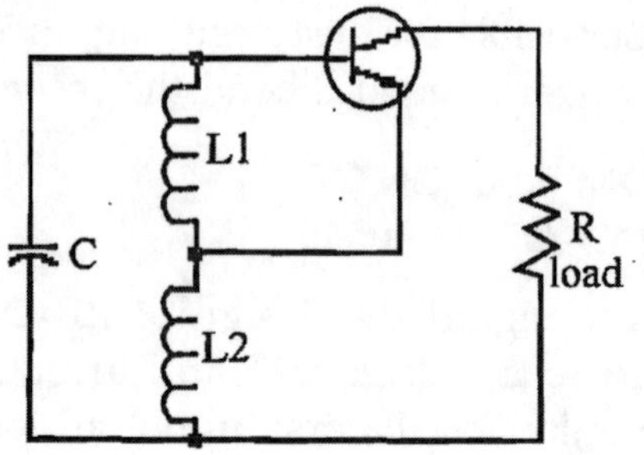

Colpitts Oscillator

The basic Colpitts oscillator circuit look like this and you will see some similarities.

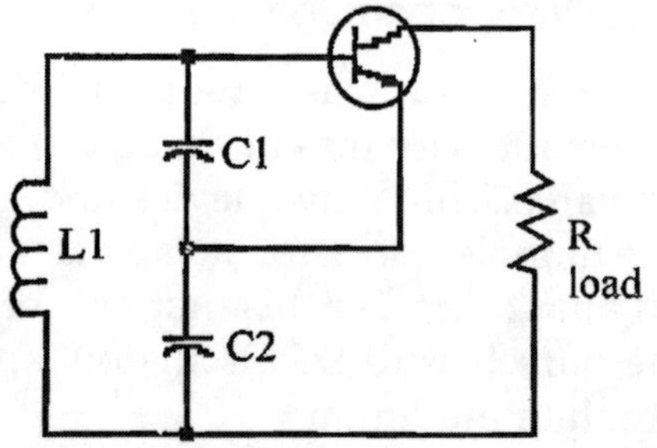

If you consider positive feedback is applied to compensate for the losses in the tuned circuit, the amplifier and feedback circuit create a negative resistor. When Z1 and Z2 are capacitive, the impedance across the capacitors can be estimated from a formula I won't lay on you here because it includes beta, hie, as well as X_{C1} and X_{C2}. Suffice to say it can be shown that the input impedance is a negative resistor in series with C1 and C2. And the frequency is in accordance with:

$$F_0 = \frac{1}{2\pi[LC_1C_2/(C_1+C_2)]^{1/2}}$$

Frequency or Phase Stability

Frequency or phase stability of an oscillator is customarily considered in the long term stability case where frequency changes are measured over minutes, hours, days even years. Of interest here are the effects of the components changes, with ambient conditions, on the frequency of oscillation. These might be caused by changes in the input voltage, variations in temperature, humidity and ageing of our components.

Never underestimate the effects of these variations on the frequency of operation. I've gone nuts working on so called precision designs, with precision components, where the frequency wandered at random over several kilohertz over several minutes. Needless to say I'd "messed up".

Short term stability is also of great interest and, again I could lay some real heavy maths on you but I won't. I'll simply say it can be mathematically proven that the higher the circuit Q, the higher this stability factor becomes. The higher the circuit Q, the better the ability the tuned circuit can filter out undesired harmonics AND noise.

Reducing Phase Noise

1. Maximize the Qu of the resonator.
2. Maximize reactive energy by means of a high RF voltage across the resonator. Use a low LC ratio.
3. Avoid device saturation and try to use anti parallel (back to back) tuning diodes.
4. Choose your active device with the lowest NF.
5. Choose a device with low flicker noise, this can be reduced by RF feedback. A bipolar transistor with an unby-passed emitter resistor of 10 to 30 ohms can improve flicker noise by as much as 40 dB.
6. The output circuits should be isolated from the oscillator circuit and take as little power as possible.

Effects of Ambient Changes on Stability

A frequency change of a few tens of hertz back and forth over a couple of minutes would mean nothing to an entertainment receiver designed for the FM Radio band. Such a drift in an otherwise contest grade receiver designed to receive CW (morse code) would be intolerable. It's a question of relativity.

Minimizing Frequency Drift

These are random and not in any particular order.

1. Isolate the oscillator from succeeding stages with a well designed buffer stage followed by a stage of amplification. Large signals can often then be reduced by a 3 or 6 dB attenuator which also has the benefit of presenting a well defined load impedance to the amplifier. If the stage is feeding a mixer, as is most often the case, then another benefit is the mixer (you are using double balanced mixers?), also see a source impedance of 50 ohms.
2. Ensure the mechanical stability of your oscillator is such that mechanical vibration can have no effect on components, especially those frequency determining components.
3. Supply the oscillator with a clean well regulated supply. If using varactor tuning, doubly ensure the tuning DC voltage is as clean as possible, a few hundred micro volts of noise can be imposed on the oscillator signal. Use back to back diodes for the variable element. Air variables are hard to come by although they offer far superior Q figures. DC tuning tends to be more versatile.
4. Minimize circuit changes from ambient variations by using NPO capacitors, polystyrene are dearer but excellent, silvered mica in my opinion are not what many people believe and are highly over rated.
5. The inductor should be air wound on a coil form with a configuration to maximize Qu. If you must use a toroid, where possible try to use the 6 type as it offers the best Q. Sometimes, for other reasons you might have to use a slug tuned form.
6. Parallel a number of smaller value NPO capacitors rather than using one large one in frequency determining components. For trimmers try and use an air variable. Keep an eye out for small value N750,

N1500 capacitors, < 15 pF, when available and are found to be dirt cheap. These are sometimes useful in taming drift in an oscillator.

7. Bipolar or FETS for active device seems to be a matter of personal preference and I've seen some ferocious arguments over that one. Consensus seems to come down in favour of FETS. Me, I'm a bipolar man because FETS hate me pure and simple.

So there are some of the things to keep in mind. Hopefully at this point you have discovered the broad idea of an oscillator, I've outlined broadly two types Hartley and Colpitts. I spoken about frequency stability and listed ways to combat phase noise and reducing frequency drift. Now let's proceed to the main course.

Designing a Hartley Oscillator

Here I'll present the schematic for my old favourite, together with a buffer stage and an amplifier stage which should deliver about 5V P/P into a 50 ohm load. We'll discuss each relevant stage and produce some rule-of-thumb design info. Because the consensus comes down in favour of FETS and I'm big enough to lay aside my prejudices in the noble cause of advanced education we'll use a FET oscillator. Nothing to do with a few friends who might belt me up!

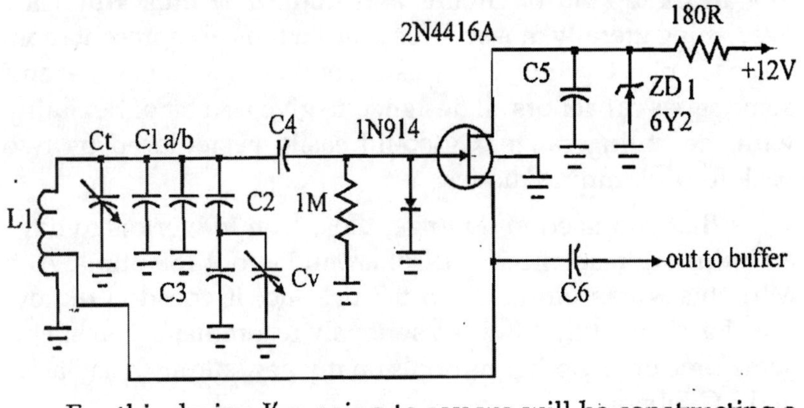

For this design I'm going to say we will be constructing a general purpose VFO to operate at 5000 - 5100 Khz no particular reason, pick anything you like.

Now I chose a 2N4416A FET purely because I bought a big bag of them years ago and have them on hand. You could use any general purpose JFET you can readily obtain. Note the 2N4416A is a metal can and the case is grounded.

The frequency determining components are L1, Ct (a nominal 10 pf trimmer), C1a, C1b, C2, C3, Cv and C4.

Note: I have been asked a number of times the function of C4 in this circuit. Capacitor C4 is to reduce the loading on the tuned circuit components. It may be as small as possible consistent with being able to provide sufficient drive to the succeeding buffer amplifier stage. Often the home constructor will often make C4 a trimmer.

The other components are bog standard. The two resistors, silicon diode and zener diode need never change, capacitor C5 is about right for this frequency. C6 can be selected to give higher / lower output to the buffer amplifier. Smaller C6 values give lower output and conversely higher values give larger output.

The silicon diode I'll explain later, the zener diode is to give a regulated 6.2 volt supply

Now there is NOTHING sacred about my frequency determining capacitor combination O.K.? Too many people look at these kind of circuits and think they must duplicate everything literally, not so. This is just a typical representation. C1 to C3 plus Cv and Ct are just a combination of parallel and some series capacitors all designed to give us a bit of flexibility with the tuning range. Cv could easily be replaced by two back to back tuning diodes.

What you need to do to get the circuit to work is to have an inductive reactance for L1 of around about 180 ohms. At 5 Mhz this works out at about 5.7 uH and, if you don't know how I arrived at that figure I seriously recommend you spend some time on my other tutorials on my new site such as Basics and LC Filters.

The important aspect is that the feedback point from the source of the JFET connects to about 25% of the windings of

L1 from the ground end. Now I've depicted an air cored inductor. It could be, just as one example among a great many, 18 - 19 turns of #20 gauge wire on a 25.4 mm (1") diameter form spread evenly over a length of about 25.4 mm (1"). The tap would be at about 4 1/2 turns. Check that out with the formula's I taught you elsewhere.

Alternatively, with degraded performance, you could use a T50-6 toroid and wind say 37 turns of #24 wire (5.48 uH) tapping at 9 turns. The A_L factor for a T50-6 is 40. Again do the other tutorials if necessary, I'm not going to repeat old work and it's going to be even harder from here on. I'll thoroughly explain new concepts, not the old.

So if we are to have our oscillator working at about 5 Mhz, we know the LC is 1013 and if L is say 5.7 uH then total C for resonance (just like LC Filters eh!) is about 177 pF. We want to be able to tune from 5000 to 5100 Khz a tuning ratio of 1.02 which means a capacitance ratio of 1.04 (min to max.).

Let's fiddle with some numbers! I have a Jackson Bros. air variable capacitor (very Rolls-Royce) which swings from 10.5 pF to 105 pF, a typical 10:1 ratio in air variables. This I will use for Cv.

If the total swing is 1.04 (actually 1.0404:1) and Cmax is 177 pF it follows Cmin is 170 pF. A variation of only 7 pF (roughly). Now we're treading on unsafe ground here with such a large variable capacitor. We could:

(A) Rip plates of it to reduce capacitance (don't even think about it)

(B) Go to varactor diodes with a small swing. That's O.K. but performance becomes degraded.

(C) Obtain a smaller air variable with Cmax of say 25 pF.

Just to prove I'm a glutton for punishment and if you're still here so are you, we will purely for the mathematical exercise, persevere with the 105 pF variable. What if we eliminate C3 and make C2 = 15 pF NPO then the series combination of C2 and Cv swing 6.176 pF to 13.125 pF, a variation of over 6.9 pF - are you lost? Go back to the other tutorials.

If our Cmax was 177 pF then 177 - 13.125 = 163.875 and the 177 pF was approximate anyway. I'd make Ct a 10 pF air trimmer (if available, if not, a ceramic or whatever the supplier offers but 10 pF max.). That leaves about 154 pF to make up. How about making C1a and C1b into 3 NPO capacitors of say 2 X 47 pF and 1 X 56 pF all NPO types. In total that comes to less than 177 pF max. but don't forget there are stray capacitance's in the circuit. In the final wash-up you could simply use 3 X 47 pF.

Fourier transform

In mathematics, the Fourier transform, named in honour of French mathematician Joseph Fourier, is a certain linear operator that maps functions to other functions. Loosely speaking, the Fourier transform decomposes a function into a continuous spectrum of its *frequency components*, and the inverse transform synthesizes a function from its spectrum of frequency components. A useful analogy is the relationship between a set of notes in musical notation (the frequency components) and the sound of the musical chord represented by these notes (the function/signal itself).

Using physical terminology, the Fourier transform of a signal $x(t)$ can be thought of as a representation of a signal in the "frequency domain"; i.e. how much each frequency contributes to the signal. This is similar to the basic idea of the various other Fourier transforms including the Fourier series of a periodic function. (*See also fractional Fourier transform and linear canonical transform for generalizations.*)

Definitions

There are several common conventions for defining the Fourier transform of a complex-valued Lebesgue integrable function, x. In communications and signal processing, for instance, it is often the function:

$$X(f) = \int_{-\infty}^{\infty} x(t)e^{-2\pi ft}dt, \text{ for every real number } f.$$

When the independent variable represents *time* (with SI unit of seconds), the transform variable f represents ordinary frequency (in hertz). The complex-valued function, X is said to represent in the frequency domain. I.e., if is a continuous function, then it can be reconstructed from X by the inverse transform:

$$x(t) = \int_{-\infty}^{\infty} X(f)e^{i2\pi ft}df, \text{ for every real number } t.$$

Other notations for $X(f)$ are: $\hat{x}(f)$ and $F\{x\}(f)$.

The interpretation of X is aided by expressing it in polar coordinate form: $X(f) = A(f)\, e^{i\phi(f)}$, where:

$A(f) = |X(f)|$, the amplitude

$\phi(f) = \angle X(f)$, the phase.

Then the inverse transform can be written:

$$x(t) = \int_{-\infty}^{\infty} A(f)e^{i(2\pi ft+\phi(f))}df,$$

which is a recombination of all the frequency components of $x(t)$. Each component is a complex sinusoid of the form $e^{i2\pi ft}$ whose amplitude is $A(f)$ and whose initial phase angle (at t = 0) is $\varphi(f)$.

In mathematics, the Fourier transform is commonly written in terms of angular frequency: $\omega = 2\pi f$, whose units are radians per second.

The substitution $f = \dfrac{\omega}{2\pi}$ into the formulas above produces this convention:

$$X(\omega) = \int_{-\infty}^{\infty} x(t)e^{-i\omega t}dt$$

$$x(t) = \frac{1}{2\pi}\int_{-\infty}^{\infty} X(\omega)e^{-i\omega t}\,d\omega,$$

which is also a bilateral Laplace transform evaluated at s = $i\omega$.

The 2π factor can be split evenly between the Fourier transform and the inverse, which leads to another popular convention:

$$x(\omega) = \frac{1}{2\pi}\int_{-\infty}^{\infty} x(t)e^{-i\omega t}\,dt$$

$$x(t) = \frac{1}{2\pi}\int_{-\infty}^{\infty} X(\omega)e^{-i\omega t}\,d\omega,$$

This convention and the $X(f)$ convention are unitary transforms.

Variations of all three conventions can be created by conjugating the complex-exponential kernel of both the forward and the reverse transform. The signs must be opposites. Other than that, the choice is (again) a matter of convention.

Summary of Popular Forms of the Fourier Transform

angular frequency ω unity

$$X_1(\omega) \stackrel{def}{=} \frac{1}{2\pi}\int_{-\infty}^{\infty} x(t)e^{-i\omega t}\,dt$$

(rad/s) $$= \frac{1}{\sqrt{2\pi}}X_2(\omega) = \frac{1}{\sqrt{2\pi}}X_3\left(\frac{\omega}{2\pi}\right)$$

$$x(t) = \frac{1}{\sqrt{2\pi}}\int_{-\infty}^{\infty} X_1(\omega)e^{i\omega t}\,d\omega$$

non-unitary $$x(t) = \frac{1}{2\pi}\int_{-\infty}^{\infty} X_2(\omega)e^{i\omega t}\,d\omega$$

ordinary unitary $$X_3(f) \stackrel{def}{=} \int_{-\infty}^{\infty} x(t)e^{-i2\pi ft}\,dt$$

frequency $= \sqrt{2\pi}X_1(2\pi f) = X_2(2\pi f)$

f $\quad x(t) = \int_{-\infty}^{\infty} X_3(f)e^{i2\pi ft}\,df$

(hertz)

Some Fourier transform properties

Notation: $f(t) \overset{F}{\Leftrightarrow} F(\omega)$ denotes that $f(t)$ and $F(\omega)$ are a Fourier transform pair.

Linearity

$$a.f(t) + b.g(t) \overset{F}{\Leftrightarrow} a.F(\omega)b.G(\omega)$$

Multiplication

$$f(t).g(t) \overset{F}{\Leftrightarrow} \frac{1}{\sqrt{2\pi}}.F(\omega)*G(\omega)$$

(unitary normalization convention)

$$\overset{F}{\Leftrightarrow} \frac{1}{2\pi}.F(\omega)*G(\omega)$$

(non-unitary convention)

$$\overset{F}{\Leftrightarrow} F(f)*G(f)$$

(ordinary frequency)

e.g., Modulation

$$f(t).\cos\omega_0 t \overset{F}{\Leftrightarrow} \frac{1}{2}[F(\omega+\omega_0)+F(\omega-\omega_0)],\ \omega_0 \in \mathbb{R}$$

$$f(t).\cos\omega_0 t \overset{F}{\Leftrightarrow} \frac{i}{2}[F(\omega+\omega_0)-F(\omega-\omega_0)]$$

$$f(t).e^{i\omega_0 t} \overset{F}{\Leftrightarrow} F(\omega-\omega_0)$$

Convolution

$$f(t) * g(t) \overset{F}{\Leftrightarrow} \sqrt{2\pi}.F(\omega).G(\omega) \quad \text{(unitary convention)}$$

$$\overset{F}{\Leftrightarrow} F(\omega).G(\omega) \quad \text{(non-unitary convention)}$$

$$\overset{F}{\Leftrightarrow} F(f).G(f) \quad \text{(ordinary frequency)}$$

e.g., Integration

$$f(t) * u(t) = \int_{-\infty}^{t} f(\tau)d\tau \overset{F}{\Leftrightarrow} \frac{1}{i\omega}F(\omega) + \pi F(0).\delta(\omega)$$

Conjugation

$$\overline{f(t)} \overset{F}{\Leftrightarrow} \overline{F(-\omega)}$$

Scaling

$$f(at) \overset{F}{\Leftrightarrow} \frac{1}{|a|}F\left(\frac{\omega}{a}\right), \quad a \in \mathbb{R}, a \neq 0$$

Time reversal

$$f(-t) \overset{F}{\Leftrightarrow} F-\omega$$

Time shift

$$f(t-t_0) \overset{F}{\Leftrightarrow} e^{-i\omega t_0}.F(\omega)$$

Parseval's theorem

$$\int_{-\infty}^{\infty} f(t).\overline{g(t)}\,dt = \int_{-\infty}^{\infty} F(\omega).\overline{G(\omega)}d\omega$$

(unitary convention)

$$= \frac{1}{2\pi}.\int_{-\infty}^{\infty} F(\omega).\overline{G(\omega)}d\omega \quad \text{(non-unitary convention)}$$

$$= \int_{-\infty}^{\infty} F(f).\overline{G(f)}df \quad \text{(ordinary frequency)}$$

The section "Table of important Fourier transforms" (below) documents more properties of the continuous Fourier transform.

Generalization

Using two arbitrary real constants a and b, the most general definition of the forward 1-dimensional Fourier transform is given by:

$$X(\omega)=\sqrt{\frac{|b|}{(2\pi)^{1-a}}}\int_{-\infty}^{+\infty} x(t).e^{ib\omega t}\,dt,$$

and the inverse is given by:

$$x(t)=\sqrt{\frac{|b|}{(2\pi)^{1+a}}}\int_{-\infty}^{+\infty} X(\omega).e^{ib\omega t}\,d\omega\,.$$

Note that the transform definitions are symmetric; they can be reversed by simply changing the signs of a and b.

The ordinary frequency convention corresponds to $(a,b) = (0, 2\pi)$, and in that case the variable ω is changed to f. If f and t carry units, their product must be dimensionless. For example, t may be in units of time, specifically seconds, and f would be in hertz.

The unitary, angular frequency convention is $(a,b) = (0,1)$, and the non-unitary convention (above) is $(a,b) = (1,1)$.

The "forward" and "inverse" transforms are always defined so that the operation of both transforms in either order on a function will return the original function. In other words, the composition of the transform pair is defined to be the identity transformation.

MORE PROPERTIES

Completeness

We define the Fourier transform on the set of compactly-supported complex-valued functions of R and then extend it by continuity to the Hilbert space of square-integrable functions with the usual inner-product. Then F: $L^2(R) \rightarrow L^2(R)$

is a unitary operator. That is $F^* = F^{-1}$ and the transform preserves inner-products (see Parseval's theorem, also described below). Note that, F^* refers to adjoint of the Fourier Transform operator.

Moreover we can check that:

$$F^2 = J,\ F^3 = F^* = F^{-1}, \text{ and } F^4 = I,$$

where J is the Time-Reversal operator defined as:

$$||J\{f\}(t) - f(-t)||_2 = 0,$$

and J is the Identity operator defined as:

$$||J\{f\}(t) - f(t)||_2 = 0.$$

Extensions

The Fourier transform can also be extended to the space of integrable functions defined on R^n:

$$F : L^1(\mathbb{R}^n) \to C(\mathbb{R}^n),$$

where:

$$L^1(\mathbb{R}^n) = \{f : \mathbb{R}^n \to \mathbb{C} \mid \int_{\mathbb{R}^n} |f(x)|\, dx < \infty\},$$

and $C(R^n)$ is the space of continuous functions on R^n.

In this case the definition usually appears as:

$$F(\omega) \overset{def}{=} \left(\frac{1}{\sqrt{2\pi}}\right)^n \int_{\mathbb{R}^n} f(x) e^{-i(\omega . x)} dx.$$

where ω " R^n and $\omega \cdot x$ is the inner product of the two vectors ω and x.

One may now use this to define the continuous Fourier transform for compactly supported smooth functions, which are dense in $L^2(R^n)$. The Plancherel theorem then allows us to extend the definition of the Fourier transform to functions on $L^2(R^n)$ (even those not compactly supported) by continuity arguments. All the properties and formulas listed on this page apply to the Fourier transform so defined.

THE PLANCHEREL THEOREM AND PARSEVAL'S THEOREM

It should be noted that depending on the author either of these theorems might be referred to as the Plancherel theorem or as Parseval's theorem.

If $f(t)$ and $g(t)$ are square-integrable and $F(\omega)$ and $G(\omega)$ are their unitary Fourier transforms, then we have Parseval's theorem:

$$\int_{\mathbb{R}^n} f(t)\overline{g}(t)dt = \int_{\mathbb{R}^n} F(\omega)\overline{G}(\omega)d\omega,$$

where the bar denotes complex conjugation. Therefore, the Fourier transformation yields an isometric automorphism of the Hilbert space $L^2(R^n)$.

The Plancherel theorem, which is equivalent to Parseval's theorem, states:

$$\int_{\mathbb{R}^n} |f(t)|^2 = \int_{\mathbb{R}^n} |F(\omega)|^2 \, d\omega.$$

This theorem is usually interpreted as asserting the unitary property of the Fourier transform. See Pontryagin duality for a general formulation of this concept in the context of locally compact abelian groups.

Localization Property

As a rule of thumb: the more concentrated $f(t)$ is, the more spread out $F(\omega)$ is. In particular, if we "squeeze" a function in t, it spreads out in ω and vice-versa; and we cannot arbitrarily concentrate both the function and its Fourier transform.

Therefore a function which equals its Fourier transform strikes a precise balance between being concentrated and being spread out. It is easy in theory to construct examples of such functions (called self-dual functions) because the Fourier transform has order 4 (that is, iterating it four times on a function returns the original function). The sum of the four iterated Fourier transforms of any function will be self-dual. There are also some explicit examples of self-dual functions, the most important being constant multiples of the Gaussian function

$$f(t) = \exp\left(\frac{-t^2}{2}\right).$$

This function is related to Gaussian distributions, and in fact, is an eigenfunction of the Fourier transform operators. Again, it is worth stressing that the mere fact that the Gaussian is self-dual does not make it in any way special: many self-dual functions exist.

The trade-off between the compaction of a function and its Fourier transform can be formalized in the form of a Fourier Uncertainty Principle. Suppose $f(t)$ and $F(\omega)$ are a Fourier transform pair for a finite-energy (i.e. square-integrable) function. Without loss of generality, we assume that $f(t)$ is normalized:

$$\int_{-\infty}^{\infty} |f(t)|^2 \, dt = 1.$$

It follows from Parseval's theorem that $F(\omega)$ is also normalized.

Define the expected *location* of a particle (with probability density $|f(t)|^2$) as

$$u_f \overset{def}{=} \int_{-\infty}^{\infty} t\,|f(t)|^2 \, dt.$$

and the expectation value of the *momentum* of the particle (with probability density $|f(\omega)|^2$) as

$$\xi_F \overset{def}{=} \int_{-\infty}^{\infty} \omega\,|F(\omega)|^2 \, d\omega.$$

Also define the variances around the above-defined average values as

$$\sigma_f^2 \overset{def}{=} \int_{-\infty}^{\infty} (t - u_f)^2\,|f(t)|^2 \, dt$$

and

$$\sigma_F^2 \overset{def}{=} \int_{-\infty}^{\infty} (\omega - \xi_F)^2\,|F(\omega)|^2 \, d\omega.$$

Then it can be shown that

$$\sigma_f^2 \sigma_F^2 \geq \frac{1}{4}.$$

The equality is achieved for the Gaussian function listed above, which shows that the Gaussian function is maximally concentrated in "time-frequency".

The most famous practical application of this property is found in quantum mechanics. Following from the axioms of quantum mechanics, the momentum and position wave functions are Fourier transform pairs to within a factor of $h/2\pi$ and are normalized to unity. The above expression then becomes a statement of the Heisenberg uncertainty principle.

The Fourier transform also translates between smoothness and decay. If $f(t)$ is several times differentiable, then $F(\omega)$ decays rapidly towards zero for $\omega \to \pm\infty$.

Analysis of Differential Equations

Fourier transforms, and the closely related Laplace transforms are widely used in solving differential equations. The Fourier transform is compatible with differentiation in the following sense: if $f(t)$ is a differentiable function with Fourier transform $F(\omega)$, then the Fourier transform of its derivative is given by $i\omega\, F(\omega)$. This can be used to transform differential equations into algebraic equations. Note that this technique only applies to problems whose domain is the whole set of real numbers. By extending the Fourier transform to functions of several variables (as outlined below), partial differential equations with domain R^n can also be translated into algebraic equations.

Convolution Theorem

The Fourier transform translates between convolution and multiplication of functions. If $f(t)$ and $h(t)$ are integrable functions with Fourier transforms $F(\omega)$ and $H(\omega)$ respectively, and if the convolution of f and h exists and is absolutely integrable, then the Fourier transform of the convolution is given by the product of the Fourier transforms $F(\omega)$ $H(\omega)$ (possibly multiplied by a constant factor depending on the Fourier normalization convention).

In the unitary normalization convention, this means that if:

$$f(t) = \{f * h\}(t) = \int_{-\infty}^{\infty} f(\tau)h(t-\tau)d\tau,$$

where * denotes the convolution operation, then:

$$G(\omega) = \sqrt{2\pi}.F(\omega)H(\omega).$$

The above formulas hold true for functions defined on both one- and multi-dimension real space. In linear time invariant (LTI) system theory, it is common to interpret $h(t)$ as the impulse response of an LTI system with input $f(t)$ and output $g(t)$, since substituting the unit impulse for $f(t)$ yields $g(t)=h(t)$. In this case, $H(\omega)$ represents the frequency response of the system.

Conversely, if $f(t)$ can be decomposed as the product of two other functions $p(t)$ and $q(t)$ such that their product $p(t)q(t)$ is integrable, then the Fourier transform of this product is given by the convolution of the respective Fourier transforms $P(\omega)$ and $Q(\omega)$, again with a constant scaling factor.

In the unitary normalization convention, this means that if $f(t) = p(t)\ q(t)$ then:

$$F(\omega) = \frac{1}{\sqrt{2\pi}}(P(\omega) * Q(\omega)) = \frac{1}{\sqrt{2\pi}} \int_{-\infty}^{\infty} P(\alpha)Q(\omega-\alpha)d\alpha.$$

Cross-correlation Theorem

In an analogous manner, it can be shown that if $g(t)$ is the cross-correlation of $f(t)$ and $h(t)$:

$$g(t) = (f * h)\ (t) = \int_{-\infty}^{\infty} \overline{f}(\tau)h(t+\tau)d\tau$$

then the Fourier transform of $g(t)$ is:

$$G(\omega) = \sqrt{2\pi}\,\overline{F}(\omega)H(\omega)$$

where capital letters are again used to denote the Fourier transform.

Tempered Distributions

The most general and useful context for studying the continuous Fourier transform is given by the tempered distributions; these include all the integrable functions mentioned above and have the added advantage that the Fourier transform of any tempered distribution is again a tempered distribution and the rule for the inverse of the Fourier transform is universally valid. Furthermore, the useful Dirac delta is a tempered distribution but not a function; its Fourier transform is a constant function (whose specific value depends upon the form of the Fourier transform used). Distributions can be differentiated and the above mentioned compatibility of the Fourier transform with differentiation and convolution remains true for tempered distributions.

Fourier Series

The Fourier series is a mathematical tool used for analyzing periodic functions by decomposing such a function into a weighted sum of much simpler sinusoidal component functions sometimes referred to as normal Fourier modes, or simply modes for short. The weights, or coefficients, of the modes, are a one-to-one mapping of the original function. Generalizations include generalized Fourier series and other expansions over orthonormal bases.

Fourier series serve many useful purposes, as manipulation and conceptualization of the modal coefficients are often easier than with the original function. Areas of application include electrical engineering, vibration analysis, acoustics, optics, signal and image processing, and data compression. Using the tools and techniques of spectroscopy, for example, astronomers can deduce the chemical composition of a star by analyzing the frequency components, or spectrum, of the star's emitted light. Similarly, engineers can optimize the design of a telecommunications system using information about the spectral components of the data signal that the system will carry (see also spectrum analyser).

The Fourier series is named after the French scientist and mathematician Joseph Fourier, who used them in his influential work on heat conduction.

DEFINITION

General Form

Given a complex-valued function f, of real argument t, f: $R \to C$, where $f(t)$ is piecewise smooth and continuous, periodic with period T, and square-integrable over the interval from t_1 to t_2 of length T, that is,

$$\int_{t_1}^{t_2} |f(t)|^2 \, dt < +\infty$$

where

$T = t_2 - t_1$ is the period,

t_1 and t_2 are integration bounds.

The Fourier series expansion of f is

$$f(t) = \frac{1}{2}a_0 + \sum_{n=1}^{\infty}[a_n \cos(\omega_n t) + b_n \sin(\omega_n t)]$$

where, for any non-negative integer n,

$\omega_n = n\frac{2\pi}{T}$ is the nth harmonic (in radians) of the function f,

$a_n = \frac{2}{T}\int_{t_1}^{t_2} f(t)\cos(\omega_n t)dt$ are the even Fourier coefficients of f, and

$b_n = \frac{2}{T}\int_{t_1}^{t_2} f(t)\sin(\omega_n t)dt$ are the odd Fourier coefficients of f.

Equivalently, in complex exponential form,

$$f(t) = \sum_{n=-\infty}^{+\infty} c_n e^{i\omega_n t}$$

where:

$$c_n = \frac{1}{T}\int_{t_1}^{t_2} f(t)e^{-i\omega_n t}\,dt$$

i is the imaginary unit, and

$e^{i\omega_n t} = \cos(\omega_n t) + i\sin(\omega_n t)$ in accordance with Euler's formula.

Canonical Form

In the special case where the period $T = 2\pi$, we have

$$\omega_n = n$$

In this case, the Fourier series expansion reduces to a particularly simple form:

$$f(t) = \frac{1}{2}a_0 + \sum_{n=1}^{\infty}[a_n \cos(nt) + b_n \sin(nt)]$$

where

$$a_n = \frac{1}{\pi}\int_{-\pi}^{\pi} f(t)\cos(nt)dt$$

$$b_n = \frac{1}{\pi}\int_{-\pi}^{\pi} f(t)\sin(nt)dt$$

for any non-negative integer n.

or, equivalently:

$$f(t) = \sum_{n=-\infty}^{+\infty} c_n e^{\text{int}}$$

where

$$c_n = \frac{1}{2\pi}\int_{-\pi}^{\pi} f(t)e^{-\text{int}}dt \frac{1}{2}(a_n - ib_n).$$

Choice of the Form

The form for period T can be easily derived from the canonical one with the change of variable defined by $x = \frac{2\pi}{T}t$.

Therefore, both formulations are equivalent. However, the form for period T is used in most practical cases because it is directly applicable. For the theory, the canonical form is preferred because it is more elegant and easier to interpret mathematically, as will later be seen.

Simple Fourier Series

Let f be periodic of period 2π, with $f(x) = x$ for x from $-\pi$ to π. Note that this function is a periodic version of the identity function.

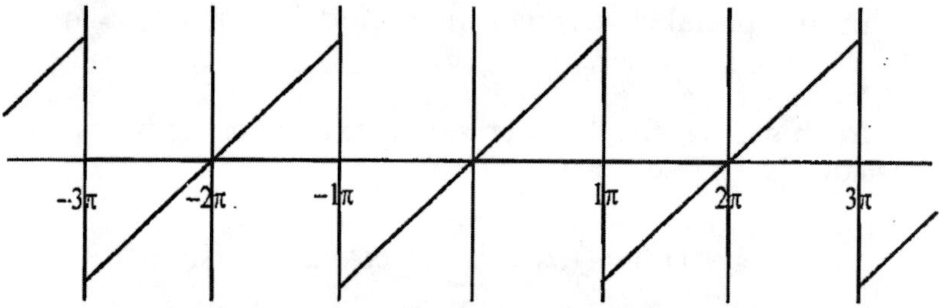

Plot of a periodic identity function - a sawtooth wave.

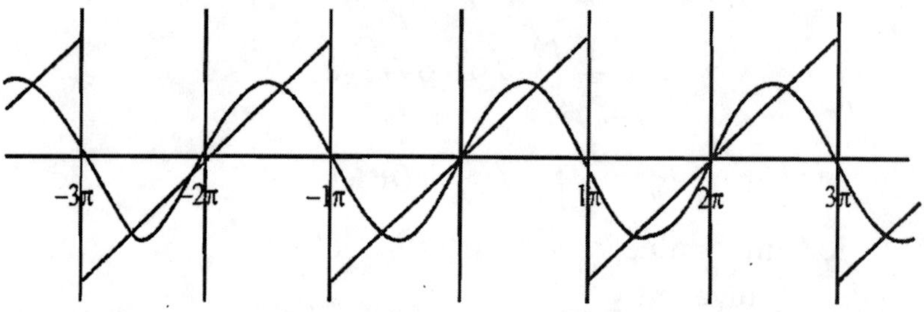

Animated plot of the first five successive partial Fourier series

We will compute the Fourier coefficients for this function.

$$a_n = \frac{1}{\pi}\int_{-\pi}^{\pi} f(x)\cos(nx)dx$$

$$= \frac{1}{\pi}\int_{-\pi}^{\pi} x\cos(nx)dx$$

$$= 0.$$

$$b_n = \frac{1}{\pi}\int_{-\pi}^{\pi} f(x)\sin(nx)dx$$

$$= \frac{1}{\pi}\int_{-\pi}^{\pi} x\sin(nx)dx$$

$$= \frac{2}{\pi}\int_{0}^{\pi} x\sin(nx)dx$$

$$= \frac{2}{\pi}\left(\left[-\frac{x\cos(nx)}{n}\right]_0^{\pi} + \left[\frac{\sin(nx)}{n^2}\right]_0^{\pi}\right)$$

$$= 2\frac{(-1)^{n+1}}{n}.$$

Notice that a_n are 0 because the $x \rightarrow x\cos(nx)$ are even functions. Hence the Fourier series for this function is:

$$f(x) = \frac{a_0}{2} + \sum_{n=1}^{\infty}[a_n \cos(nx) + b_n \sin(nx)]$$

$$= 2\sum_{n=1}^{\infty}\frac{(-1)^{n+1}}{n}\sin(nx), \quad \forall x \in [-\pi, \pi].$$

One application of this Fourier series is to compute the value of the Riemann zeta function at $s = 2$; by Parseval's theorem, we have:

$$\frac{1}{2\pi}\int_{-\pi}^{\pi} x^2 dx = \frac{1}{2}\sum_{n>0}\left[2\frac{(-1)^n}{n}\right]^2$$

which yields: $\sum_{n>0}\frac{1}{n^2} = \frac{\pi^2}{6}$.

The Wave Equation

The wave equation governs the motion of a vibrating

string, which may be fastened down at its endpoints. The solution of this problem requires the trigonometric expansion of a general function f that vanishes at the endpoints of an interval $x = 0$ and $x = L$. The Fourier series for such a function takes the form

$$f(x) = \sum_{n=1}^{\infty} b_n \sin\left(\frac{n\pi}{L}x\right)$$

where

$$b_n = \frac{2}{L}\int_0^L f(x)\sin\left(\frac{n\pi}{L}\right)dx.$$

Vibrations of air in a pipe that is open at one end and closed at the other are also described by the wave equation. Its solution requires expansion of a function that vanishes at $x = 0$ and whose derivative vanishes at $x = L$. The Fourier series for such a function takes the form

$$f(x) = \sum_{n=1}^{\infty} b_n \sin\left(\frac{(2n+1)\pi}{2L}x\right)$$

where

$$b_n = \frac{2}{L}\int_0^L f(x)\sin\left(\frac{(2n+1)\pi}{2L}x\right)dx.$$

INTERPRETATION: DECOMPOSING A MOVEMENT IN ROTATIONS

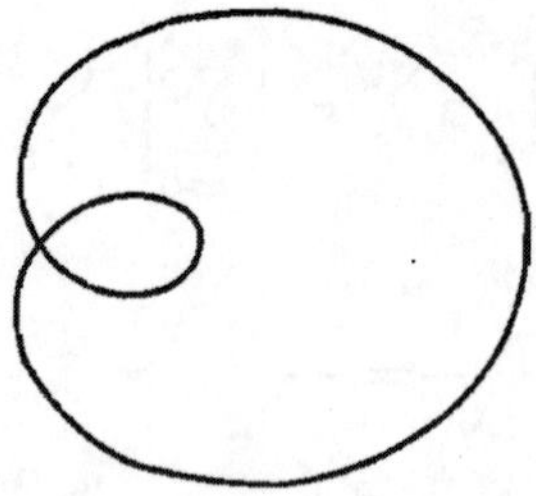

Movement in the Complex Plane

Fourier series have a kinematic interpretation. Indeed, the function $t \to f(t)$ can be seen as the movement of an object on a plane (t would then represent time). Since f is complex-valued, we can write

$$f(t) = u(t) + iv(t).$$

for real-valued functions u and v. In this form, we can interpret f as a sum of horizontal and vertical translations.

From time t to time $t + dt$, where dt is a very small incremental period, the object moves from the point $A = \begin{bmatrix} u(t) \\ v(t) \end{bmatrix}$ to the point $B = \begin{bmatrix} u(t+dt) \\ v(t+dt) \end{bmatrix}$, which corresponds to an infinitesimal translation in space by the vector $\overrightarrow{AB} = \begin{bmatrix} u(t+dt) - u(t) \\ v(t+dt) - v(t) \end{bmatrix}$. As a result, we can write f as:

$$= \int_0^t \frac{1}{dx} \begin{bmatrix} u(t+dt) - u(t) \\ v(t+dt) - v(t) \end{bmatrix} dx.$$

Now instead of seeing f as a sum of infinitesimal translations, we can see it as an infinite sum of rotations of different radii. This interpretation is convenient, in particular when the movement is periodic.

Let $\chi_n = e^{inx}$ be the n-turn per second rotation (of radius 1) (sometimes called character). We want to write f as $f(x) = \sum c_n \chi_n$. We can prove (see mathematical derivation below) that the radii of the rotations (the coefficients c_n) are exactly the ones we gave in the previous paragraph.

For example, the plot of the function $f : t \to 2\cos\left(\frac{t}{2}\right)e^{\frac{3}{2}it}$ is closed, which means the function is periodic. The loop in the curve suggests that it is the sum of two periodic functions, one having a shorter period than the other. Indeed, it can be written: $f(t) = e^{it} + e^{2it} = \chi_1(t) + \chi_2(t)$. All its Fourier coefficients are zero except $c_1 = 1$ and $c_2 = 1$. The graphical interpretation of a rotation is much harder to do than that of the translations because instead of visually seeing the movement from one point to another we have to add the whole motion for the decomposition to make sense (we are reasoning in rotation frequencies rather than in time).

Mathematically, adopting this point of view is seeing Fourier series as a tool to understand linear operators that commute with translations. The functions χ_n are precisely the multiplicative characters of the group $\mathbb{R}/2\pi\mathbb{Z}$.

Historical Development

Fourier series are named in honour of Joseph Fourier, who made important contributions to the study of trigonometric series, after preliminary investigations by Madhava, Nilakantha Somayaji, Jyesthadeva, Leonhard Euler, Jean le Rond d'Alembert, and Daniel Bernoulli. He applied this technique to find the solution of the heat equation, publishing his initial results in 1807 and 1811, and publishing his *Théorie analytique de la chaleur* in 1822.

From a modern point of view, Fourier's results are somewhat informal, due to the lack of a precise notion of *function* and *integral* in the early nineteenth century (for example, one wondered if a function defined on two intervals with two different formulas was still a function). Later, Dirichlet and Riemann expressed Fourier's results with greater precision and formality.

A Revolutionary Article

In Fourier's work entitled *Mémoire sur la propagation de la chaleur dans les corps solides,* on pages 218 and 219, we can read the following :

$$\varphi(y) = a\cos\frac{xy}{2} + a'\cos 3\frac{xy}{2} + a''\cos 5\frac{xy}{2} + ...$$

Multiplying both sides by $\cos(2i+1)\frac{xy}{2}$, and then integrating from $y = -1$ to $y = +1$ yields:

$$a_i = \int_{-1}^{1} \varphi(y)\cos(2i+1)\frac{xy}{2}dy$$

In these few lines, which are surprisingly close to the modern formalism used in Fourier series, Fourier unwittingly revolutionized both mathematics and physics. Although similar trigonometric series were previously used by Euler, d'Alembert, Daniel Bernoulli and Gauss, Fourier was the first to recognize that such trigonometric series could represent *arbitrary* functions, even those with discontinuities. It has required many years to clarify this insight, and it has led to important theories of convergence, function space, and harmonic analysis.

The originality of this work was such that when Fourier submitted his paper in 1807, the committee (composed of no lesser mathematicians than Lagrange, Laplace, Malus and Legendre, among others) concluded:...*the manner in which the author arrives at these equations is not exempt of difficulties and that his analysis to integrate them still leaves something to be desired on the score of generality and even rigour.*

The Birth of Harmonic Analysis

Since Fourier's time, many different approaches to defining and understanding the concept of Fourier series have been discovered, all of which are mathematically equivalent (and correct), but each of which emphasizes different aspects of the topic. Some of the more powerful and elegant approaches are based on mathematical ideas and tools that were not available at the time Fourier completed his original work. Fourier originally defined the Fourier series for real-valued functions of real arguments, and using the sine and cosine functions as the basis set for the decomposition.

Many other Fourier-related transforms have since been defined, extending to other applications the initial idea of representing any periodic function as a superposition of harmonics. This general area of inquiry is now sometimes called harmonic analysis.

Modern Derivation of the Fourier Coefficients

The method used by Fourier to derive the coefficients of the series is very practical and well-suited to the problem he was dealing with (heat propagation). However, this method has since been generalized to a much wider class of problems: writing a function as a sum of periodic functions.

More precisely, if $f: R \to C$ is a function, we would like to write this function as a sum of trigonometric functions, i.e. $f(x) = \sum c_n e^{inx}$. We have to restrict our choice of functions in order for this to make sense. First of all, if f has period T, then by changing variables, one can study $x \to f\left(\frac{T}{2\pi}x\right)$ which has period 2π. This simplifies notations a lot and allows us to use a canonical (standard) form. We can restrict the study of $x \to f\left(\frac{T}{2\pi}x\right)$ to any interval of length 2π, [-π,π], say.

We will take the functions $f: R \to C$ in the set of piecewise continuous, 2π periodic functions with $\int_{-\pi}^{\pi} |f(x)|^2 \, dx < +\infty$. Technically speaking, we are in fact taking functions from the Lp space $L^2(\mu)$, where μ is the normalized Lebesgue measure of the interval [-π,π] (i.e. such that $\int_{-\pi,\pi} f d\mu = \frac{1}{\pi}\int_{-\pi}^{\pi} f(x)dx$.

Complex Fourier Coefficients

We can make $L^2(\mu)$ into a Hilbert space, which is well-suited for orthogonal projections, by defining the inner product:

$$\langle f, g\rangle = \int_{[-\pi,\pi]} f\bar{g}\,d\mu = \frac{1}{2\pi}\int_{-\pi}^{\pi} f(x)\overline{g(x)}\,dx,$$

where $\overline{f(x)}$ denotes the conjugate of $f(x)$. We will denote by $\|.\|$ the associated norm.

$E = \{t \rightarrow e^{int}, n \in \mathbb{Z}\}$ is an orthonormal basis of $L^2(\mu)$, which means we can write

$$f(x) = \sum_{n\in\mathbb{Z}} \langle f, e^{inx}\rangle e^{inx}.$$

We usually define $\forall n \in \mathbb{Z}, c_n = \langle f, e^{inx}\rangle$. These numbers are called complex Fourier coefficients. Their expression is

$$c_n = \frac{1}{2\pi}\int_{-\pi}^{\pi} f(x)e^{inx}\,dx.$$

An equivalent formulation is to write f as a sum of sine and cosine functions.

Real Fourier Coefficients

The sum in the previous section is symmetrical around 0: indeed, except for $n = 0$, a c_{-n} coefficient corresponds to every c_n coefficient. This reminds one of the formulae

$$\cos x = \frac{e^{ix} + e^{-ix}}{2} \text{ and } \sin x = \frac{e^{ix} - e^{-ix}}{2i}.$$

It is therefore possible to express the Fourier series with real-valued functions. To do this, we first notice that

$$f(x) = \sum_{n\in\mathbb{Z}} c_n e^{inx} = c_0 + \sum_{n>0} [c_{-n}e^{-inx} + c_n e^{inx}].$$

After replacing c_n by its expression and simplifying the result we get

$$f(x) = c_0 + \sum_{n\in\mathbb{Z}}\left[\frac{1}{\pi}\left(\int_{-\pi}^{\pi} f(t)\cos(nt)dt\right)\cos(nx) + \frac{1}{\pi}\left(\int_{-\pi}^{\pi} f(t)\sin(nt)dt\right)\sin(nx)\right].$$

If, for a non-negative integer n, we define the real Fourier coefficients a_n and b_n by

$$a_n = \frac{1}{\pi}\int_{-\pi}^{\pi} f(x)\cos(nx)dx,$$

$$b_n = \frac{1}{\pi}\int_{-\pi}^{\pi} f(x)\cos(nx)dx,$$

we get:

$$f(x) = \frac{a_0}{2} + \sum_{n>0}[a_n\cos(nx) + b_n\sin(nx)].$$

Properties

The following properties can be easily derived from Euler's formula: $e^{ix} = \cos x + i\sin x$.

$a_n = c_n + c_{-n}$ and $b_n = i(c_n - n_{-n})$ for all n and

$c_n = \dfrac{a_n - ib_n}{2}$ and $c_{-n} = \dfrac{a_n + ib_n}{2}$ for all n.

If f is an odd function, then $a_n = 0$ for all n because $f(x)\cos\left(n\pi\dfrac{x}{T}\right)$ is then also odd, so its integral on [™ T,T] is zero. If f is an even function, then $b_n = 0$ for a similar reason.

If f is piecewise continuous, $\lim_{n\to+\infty} c_n(f) = 0$, $\lim_{n\to+\infty} c_{-n}(f) = 0$, $\lim_{n\to+\infty} a_n(f) = 0$ and $\lim_{n\to+\infty} b_n(f) = 0$.

If f is k-times piecewise continuously differentiable, then we can easily compute the Fourier coefficients of $f^{(k)}$ given those of f:

$$c_n(f^{(k)}) = (in)^k c_n(f),$$

where $f^{(k)}$ denotes the kth derivative of f.

For any positive integer k, if f is C^{k-1} and piecewise C^k, then

$$\lim_{n\to+\infty} |n^k C_n(f)| = 0 \text{ because } n^k C_n(f) = i^k C_n(f^{(k)}) \to 0.$$

This means that the sequence $c_n(f)$ is rapidly decreasing.

General Case

Fourier series take advantage of the periodicity of a function f but what if f is periodic in more than one variable, or for that matter, what if f is not periodic? These problems led mathematicians and theoretical physicists to try to define Fourier series on any group G. The advantage of this is that it allows us, for example, to define Fourier series for functions of several variables. Fourier series and Fourier transforms usually used in signal processing then become special cases of this theory and are easier to interpret.

If G is a locally compact Abelian group and T is the unit circle, we can define the dual of G by $\hat{G}\{\chi : G \to T \text{ homomorphism}\}$. This is the set of rotations on the unit circle and its elements are called characters. We can define a scalar product $\langle .,. \rangle$ on C[G] by: $\langle \chi_1, \chi_2 \rangle = \int_G \chi_1(g)\overline{\chi_2(g)}dg$. $\hat{G}$ is then an orthonormal basis of C[G] with respect to this scalar product. Let $f: G \to C$. The Fourier coefficients of f are defined by: $\hat{f}(\chi) = \langle f, \chi \rangle$ and we have $f(g) = \int_{\hat{G}} (\chi)\chi(g)d\chi$. If the group is discrete, then the integral reduces to an ordinary sum.

For example, the Fourier of this article are obtained by taking $G = R/2\pi Z$. We get

$$\hat{G} = \{\chi_n : t \to e^{int}, n \in \mathbb{Z}\}$$

and

$$C_n(f) = \hat{f}(\chi_n) = \int_G f(g)\overline{\chi(g)}\,dg = \frac{1}{2\pi} f(t)e^{-int}\,dt.$$

Periodic functions in n dimensions can be defined on an n-dimensional torus (the function taking a value at each point on the torus). Such a torus is defined by $T^n = R^n/(2\pi Z)^n$. For $n = 1$ we get a circle, for $n = 2$ the Cartesian product of two circles, i.e. a torus in the usual sense. Choosing $G = T^n$ gives the corresponding Fourier series.

Approximation and Convergence of Fourier Series

Definition of a Fourier Series

Let $\chi_n(x) = e^{in\pi\frac{x}{T}}$. We call Fourier series of the function f the series $\sum C_n \chi_n$. For any positive integer N, we $f_N(x) = \sum_{n=-N}^{N} C_n \chi_n(x)$ call the N-th partial sum of the Fourier series of this function.

Approximation with the Partial Sums

Say we want to find the best approximation of f using only the functions $\div_n$ for n from $-N$ to N. Let $T_N = \left\{ p = \sum_{n=-N}^{N} x_n \chi_n, x_n \in \mathbb{C} \right\}$. We are trying to find coefficients $(x_{-N}, \ldots, x_N)$ such that $\| f - p \|$ is minimum (where $\|.\|$ denotes the norm).

We have $\| f - p \|^2 = \| f \|^2 - 2\text{Re}\langle f, p \rangle + \| p \|^2$, where Re(z) denotes the real part of z.

$$\langle f, p \rangle = \sum_{n=-N}^{N} \overline{x_n} \langle f, \chi_n \rangle.$$

Parseval's theorem (which can be derived independently from Fourier series) gives us

$$\| p \|^2 = \sum_{n=-N}^{N} | x_n |^3.$$

By definition, $C_n = \langle f, \chi_n \rangle$; therefore

$$\| f - p \|^2 = \| f \|^2 + \sum_{n=-N}^{N} [| C_n - x_n |^2 - | C_n |^2].$$

It is clear that this expression is minimum for $x_n = c_n$ and for this value only.

This means that there is one and only one $F_N \in T_N$ such that

$$\| f - f_N \| = \min_{p \in T_N} \{ \| f - p \|, p \in T_N \},$$

it is given by

$$f_N(x) = \sum_{n=-N}^{N} C_n \chi_n(x),$$

where

$$C_n = \frac{1}{2T} \int_{-T}^{T} f(t) \chi_{-n}(t) dt.$$

This means that the best approximation of f we can make using only the functions $\chi_n(x) = e^{in\pi \frac{x}{T}}$ for n from - N to N is precisely the Nth partial sum of the Fourier series. An illustration of this is given on the animated plot of example.

Convergence

While the Fourier coefficients a_n and b_n can be formally defined for any function for which the integrals make sense, whether the series so defined actually converges to $f(x)$ depends on the properties of f.

The simplest answer is that if f is square-integrable then

$$\lim_{N\to\infty}\int_{-\pi}^{\pi}\left|f(x)-\sum_{n=-N}^{N}C_n\chi_n(x)\right|^2 dx=0.$$

This is convergence in the norm of the space L^2. The proof of this result is simple, unlike Lennart Carleson's much stronger result that the series actually converges almost everywhere.

There are many known tests that ensure that the series converges at a given point x, for example, if the function is differentiable at x. Even a jump discontinuity does not pose a problem: if the function has left and right derivatives at x, then the Fourier series will converge to the average of the left and right limits (but see Gibbs phenomenon). However, a fact that many find surprising, is that the Fourier series of a continuous function need not converge pointwise.

This unpleasant situation is counter-balanced by a theorem by Dirichlet which states that if f is $2T$-periodic and piecewise continuously differentiable function, then its Fourier series converges pointwise and $\sum_{n\in\mathbb{Z}}C_n\chi_n(x)=\dfrac{f(x^+)+f(x^-)}{2}$,

where $f(x^+)=\lim_{t\to x,t>x}f(x)$ and $f(x^-)=\lim_{t\to x,t>x}f(x)$. If f is continuous as well as piecewise continuously differentiable, then the Fourier series converges uniformly.

In 1922, Andrey Kolmogorov published an article entitled Une série de Fourier-Lebesgue divergente presque partout in which he gave an example of a Lebesgue-integrable function whose Fourier series diverges almost everywhere. This function is not in $L^2(\mu)$.

Plancherel's and Parseval's Theorems

Another important property of the Fourier series is the Plancherel theorem. Let $f, g \in L^2(\mu)$ and $c_n(f), c_n(g)$ be the corresponding complex Fourier coefficients. Then

$$\sum_{n\in\mathbb{Z}} C_n(f)\overline{C_n(g)} = \frac{1}{2T}\int_{-T}^{T} f(x)\overline{g(x)}\,dx$$

where $\overline{z}$ denotes the conjugate of z.

Parseval's theorem, a special case of the Plancherel theorem, states that:

$$\sum_{n\in\mathbb{Z}} |C_n(f)|^2 = \frac{1}{2T}\int_{-T}^{T} |f(x)|^2\,dx$$

which can be restated with the real Fourier coefficients:

$$\frac{a_0^2}{4} + \frac{1}{2}\sum_{n=1}^{\infty}\left(a_n^2 + b_n^2\right) = \frac{1}{2T}\int_{-T}^{T} |f(x)|^2\,dx.$$

These theorems may be proven using the orthogonality relationships. They can be interpreted physically by saying that writing a signal as a Fourier series does not change its energy.

4

Classical Electromagnetic Theory

BIVECTORS

A bivector is a two-dimensional analog to a one-dimensional vector. Whereas a vector is often utilized to represent a one-dimensional directed quantity (often visualized geometrically as a directed line-segment), a bivector is used to represent a two-dimensional directed quantity (often visualized as an oriented plane-segment).

Since a bivector is a two-dimensional entity, it can be built up from two linearly independent vectors, a and b by means of the exterior product.

$$B = a \wedge b$$

The vectors a and b span the subspace represented by the bivector B. Typically the orientation of the bivector is established by placing the two vectors tail-to-tail and sweeping from the first vector to the second. In this way, an oppositely oriented bivector may be obtained by reversing the order of the vectors in the exterior product.

$$a \wedge b = -b \wedge a$$

Maths - Bivectors

The vector cross gives a bivector rather than a vector.

Its not very strange that the product of two vectors is not a vector itself, its the same principle as the dot product which produces a scalar. In fact the cross product and dot products are complimentary.

In 3 dimensions a bivector behaves like a vector therefore, in most cases, we treat the result of cross product as a vector.

However there are some advantages in keeping track as to whether such quantities are vectors or bivectors:

- In physics, it can give us clues about the nature of a quantity, bivectors represent quantities associated with planes including rotations.
- There are some small differences between the algebra of bivectors, the dot product of two bivectors is a negative number whereas the dot product of two vectors is a positive number.
- This allows the vector cross product to be generalised to the outer product which is defined for all dimensions of vectors.
- This is a good introduction to Clifford Algebra/ Geometric Algebra which is a more general type of algebra and gives a lot of insights

Multiplying the bivector by a vector will toggle it back to a vector, so it is like an exclusive-or operation, multiplying the same types gives a bivector multiplying different types gives a vector:

vector × vector = bivector

bivector × bivector = bivector

vector × bivector = vector

bivector × vector = vector

Vector Cross Product

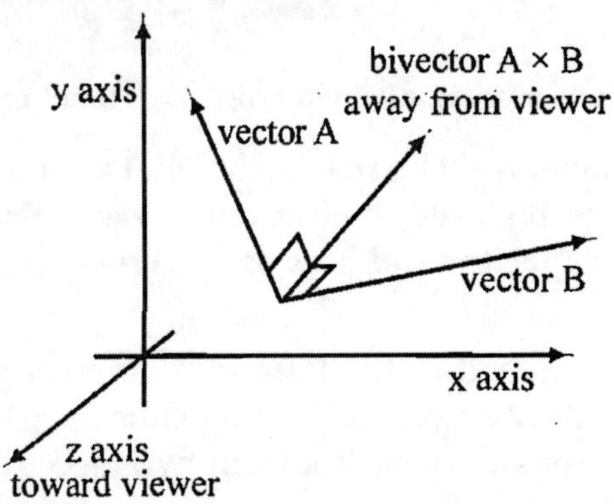

The vector cross product gives a vector (or more strictly - a bivector) which is perpendicular to both the vectors being multiplied. The resulting vector A × B is defined by:

$$x = Ay * Bz - By * Az$$
$$y = Az * Bx - Bz * Ax$$
$$z = Ax * By - Bx * Ay$$

where x, y and z are the components of A × B

CAPACITOR

A capacitor is a little like a battery. Although they work in completely different ways, capacitors and batteries both store electrical energy. If you have read How Batteries Work, then you know that a battery has two terminals. Inside the battery, chemical reactions produce electrons on one terminal and absorb electrons on the other terminal. A capacitor is much simpler than a battery, as it can't produce new electrons — it only stores them.

Fig. Flash capacitor from a point-and-shoot camera

In this article, we'll learn exactly what a capacitor is, what it does and how it's used in electronics. We'll also look at the history of the capacitor and how several people helped shape its progress.

Inside the capacitor, the terminals connect to two metal plates separated by a non-conducting substance, or dielectric. You can easily make a capacitor from two pieces of aluminum foil and a piece of paper. It won't be a particularly good capacitor in terms of its storage capacity, but it will work.

In theory, the dielectric can be any non-conductive substance. However, for practical applications, specific materials are used that best suit the capacitor's function. Mica, ceramic, cellulose, porcelain, Mylar, Teflon and even air are some of the non-conductive materials used. The dielectric dictates what kind of capacitor it is and for what it is best suited. Depending on the size and type of dielectric, some capacitors are better for high frequency uses, while some are better for high voltage applications. Capacitors can be manufactured to serve any purpose, from the smallest plastic capacitor in your calculator, to an ultra capacitor that can power a commuter bus. NASA uses glass capacitors to help wake up the space shuttle's circuitry and help deploy space probes. Here are some of the various types of capacitors and how they are used.

- Air - Often used in radio tuning circuits
- Mylar - Most commonly used for timer circuits like clocks, alarms and counters
- Glass - Good for high voltage applications
- Ceramic - Used for high frequency purposes like antennas, X-ray and MRI machines
- Super capacitor - Powers electric and hybrid cars

CAPACITOR CIRCUIT

In an electronic circuit, a capacitor is shown like this:

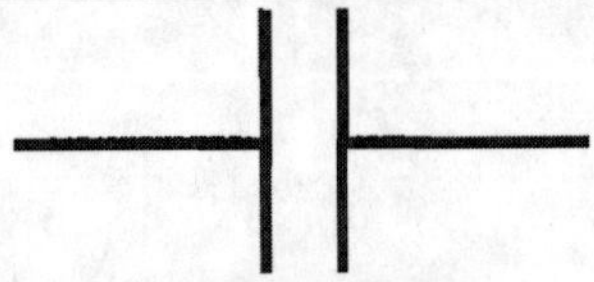

Electrical Symbol for a Capacitor

When you connect a capacitor to a battery, here's what happens:

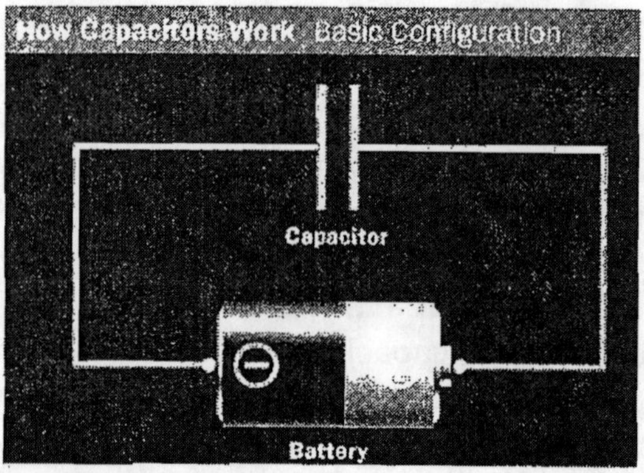

- The plate on the capacitor that attaches to the negative terminal of the battery accepts electrons that the battery is producing.
- The plate on the capacitor that attaches to the positive terminal of the battery loses electrons to the battery.

Once it's charged, the capacitor has the same voltage as the battery (1.5 volts on the battery means 1.5 volts on the capacitor). For a small capacitor, the capacity is small. But large capacitors can hold quite a bit of charge. You can find capacitors as big as soda cans that hold enough charge to light a flashlight bulb for a minute or more.

Even nature shows the capacitor at work in the form of lightning. One plate is the cloud, the other plate is the ground and the lightning is the charge releasing between these two "plates." Obviously, in a capacitor that large, you can hold a huge amount of charge!

Let's say you hook up a capacitor like this:

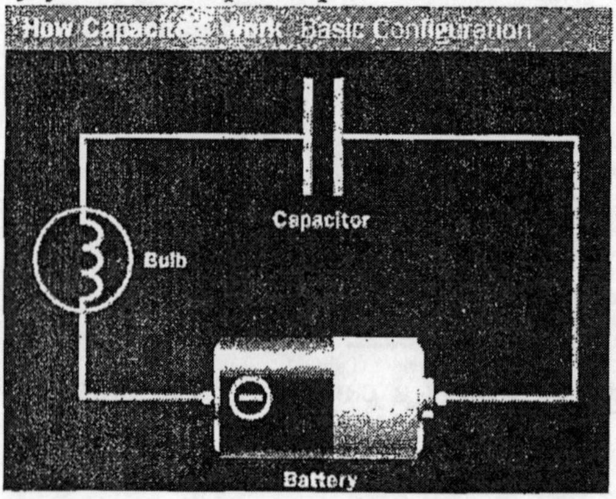

Here you have a battery, a light bulb and a capacitor. If the capacitor is pretty big, what you will notice is that, when you connect the battery, the light bulb will light up as current flows from the battery to the capacitor to charge it up. The bulb will get progressively dimmer and finally go out once the capacitor reaches its capacity. If you then remove the battery and replace it with a wire, current will flow from one plate of the capacitor to the other. The bulb will light initially and then dim as the capacitor discharges, until it is completely out.

Like a Water Tower

One way to visualize the action of a capacitor is to imagine it as a water tower hooked to a pipe. A water tower "stores" water pressure – when the water system pumps produce more water than a town needs, the excess is stored in the water tower. Then, at times of high demand, the excess water flows out of the tower to keep the pressure up. A capacitor stores electrons in the same way and can then release them later.

FARAD

A capacitor's storage potential, or capacitance, is measured in units called farads. A 1-farad capacitor can store one coulomb (coo-lomb) of charge at 1 volt. A coulomb is 6.25e18 (6.25 * 10^18, or 6.25 billion billion) electrons. One amp represents a rate of electron flow of 1 coulomb of electrons per second, so a 1-farad capacitor can hold 1 amp-second of electrons at 1 volt.

A 1-farad capacitor would typically be pretty big. It might be as big as a can of tuna or a 1-liter soda bottle, depending on the voltage it can handle. For this reason, capacitors are typically measured in microfarads (millionths of a farad).

To get some perspective on how big a farad is, think about this:

- A standard alkaline AA battery holds about 2.8 amp-hours.
- That means that a AA battery can produce 2.8 amps for an hour at 1.5 volts (about 4.2 watt-hours – a

AA battery can light a 4-watt bulb for a little more than an hour).

- Let's call it 1 volt to make the math easier. To store one AA battery's energy in a capacitor, you would need 3,600 * 2.8 = 10,080 farads to hold it, because an amp-hour is 3,600 amp-seconds.

If it takes something the size of a can of tuna to hold a farad, then 10,080 farads is going to take up a LOT more space than a single AA battery! Obviously, it's impractical to use capacitors to store any significant amount of power unless you do it at a high voltage.

Applications

The difference between a capacitor and a battery is that a capacitor can dump its entire charge in a tiny fraction of a second, where a battery would take minutes to completely discharge. That's why the electronic flash on a camera uses a capacitor — the battery charges up the flash's capacitor over several seconds, and then the capacitor dumps the full charge into the flash tube almost instantly. This can make a large, charged capacitor extremely dangerous — flash units and TVs have warnings about opening them up for this reason. They contain big capacitors that can, potentially, kill you with the charge they contain.

Capacitors are used in several different ways in electronic circuits:

- Sometimes, capacitors are used to store charge for high-speed use. That's what a flash does. Big lasers use this technique as well to get very bright, instantaneous flashes.
- Capacitors can also eliminate ripples. If a line carrying DC voltage has ripples or spikes in it, a big capacitor can even out the voltage by absorbing the peaks and filling in the valleys.
- A capacitor can block DC voltage. If you hook a small capacitor to a battery, then no current will flow between the poles of the battery once the capacitor

charges. However, any alternating current (AC) signal flows through a capacitor unimpeded. That's because the capacitor will charge and discharge as the alternating current fluctuates, making it appear that the alternating current is flowing.

Fig. A family of capacitors

In the next section, we'll look at the history of the capacitor and how some of the most brilliant minds contributed to its progress.

History of the Capacitor

The invention of the capacitor varies somewhat depending on who you ask. There are records that indicate a German scientist named Ewald Georg von Kleist invented the capacitor in November 1745. Several months later Pieter van Musschenbroek, a Dutch professor at the University of Leyden came up with a very similar device in the form of the Leyden jar, which is typically credited as the first capacitor. Since Kleist didn't have detailed records and notes, nor the notoriety of his Dutch counterpart, he's often overlooked as a contributor to the capacitor's evolution. However, over the years, both have been given equal credit as it was established that their research was independent of each other and merely a scientific coincidence

The Leyden jar was a very simple device. It consisted of a glass jar, half filled with water and lined inside and out with metal foil. The glass acted as the dielectric, although it was thought for a time that water was the key ingredient. There was usually a metal wire or chain driven through a cork in the top of the jar. The chain was then hooked to something that would deliver a charge, most likely a hand-cranked static generator. Once delivered, the jar would hold two equal but opposite charges in equilibrium until they were connected with a wire, producing a slight spark or shock

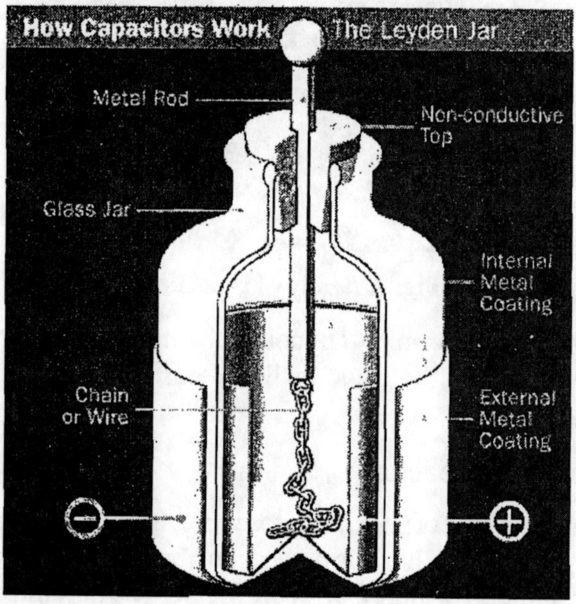

Benjamin Franklin worked with the Leyden jar in his experiments with electricity and soon found that a flat piece of glass worked as well as the jar model, prompting him to develop the flat capacitor, or Franklin square. Years later, English chemist Michael Faraday would pioneer the first practical applications for the capacitor in trying to store unused electrons from his experiments. This led to the first usable capacitor, made from large oil barrels. Faraday's progress with capacitors is what eventually enabled us to deliver electric power over great distances. As a result of Faraday's achievements in the field of electricity, the unit of measurement for capacitors, or capacitance, became known as the farad.

Capacitance

The capacitor's capacitance (*C*) is a measure of the amount of charge (*Q*) stored on each plate for a given potential difference or *voltage* (*V*) which appears between the plates:

$$C = \frac{Q}{V}$$

In SI units, a capacitor has a capacitance of one farad when one coulomb of charge is stored due to one volt applied potential difference across the plates. Since the farad is a very large unit, values of capacitors are usually expressed in microfarads (μF), nanofarads (*nF*), or picofarads (*pF*).

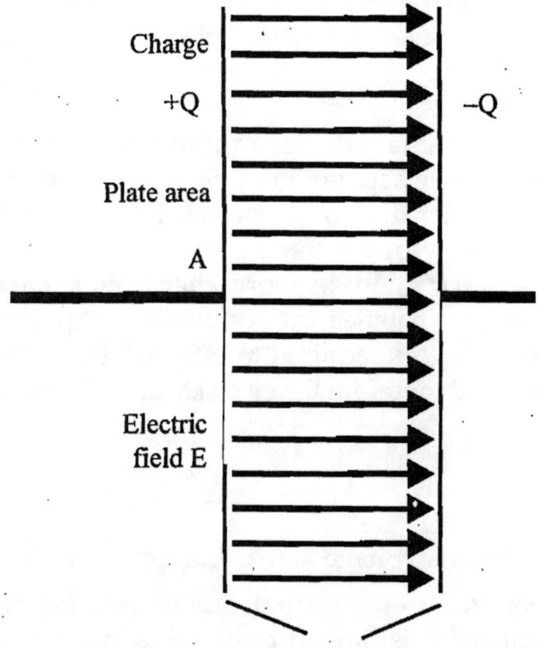

When there is a difference in electric charge between the plates, an electric field is created in the region between the plates that is proportional to the amount of charge that has been moved from one plate to the other. This electric field creates a potential difference $V = E \cdot d$ between the plates of this simple parallel-plate capacitor.

The capacitance is proportional to the surface area of the conducting plate and inversely proportional to the distance between the plates. It is also proportional to the permittivity of the dielectric (that is, non-conducting) substance that separates the plates.

The capacitance of a parallel-plate capacitor is given by:

$$C \approx \frac{\varepsilon A}{d}; A \gg d^2$$

where ε is the permittivity of the dielectric (see Dielectric constant), A is the area of the plates and d is the spacing between them.

In the diagram, the rotated molecules create an opposing electric field that partially cancels the field created by the plates, a process called dielectric polarization.

Stored Energy

As opposite charges accumulate on the plates of a capacitor due to the separation of charge, a voltage develops across the capacitor due to the electric field of these charges. Ever-increasing work must be done against this ever-increasing electric field as more charge is separated. The energy (measured in joules, in SI) stored in a capacitor is equal to the amount of work required to establish the voltage across the capacitor, and therefore the electric field. The energy stored is given by:

$$E_{\text{stored}} = \frac{1}{2}CV^2 = \frac{1}{2}\frac{Q^2}{C} = \frac{1}{2}VQ$$

where V is the voltage across the capacitor.

The maximum energy that can be (safely) stored in a particular capacitor is limited by the maximum electric field that the dielectric can withstand before it breaks down. Therefore, all capacitors made with the same dielectric have about the same maximum energy density (joules of energy per cubic meter).

Hydraulic Model

As electrical circuitry can be modeled by fluid flow, a capacitor can be modeled as a chamber with a flexible diaphragm separating the input from the output. As can be determined intuitively as well as mathematically, this provides the correct characteristics:

- The pressure difference (voltage difference) across the unit is proportional to the integral of the flow (current)
- A steady state current cannot pass through it because the pressure will build up across the diaphragm until it equally opposes the source pressure.
- But a transient pulse or alternating current can be transmitted
- The capacitance of units connected in parallel is equivalent to the sum of their individual capacitances

Electrical Circuits

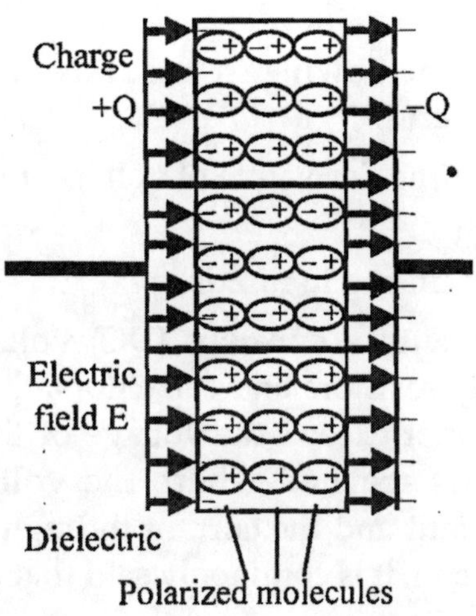

The electrons within dielectric molecules are influenced by the electric field, causing the molecules to rotate slightly from their equilibrium positions. The air gap is shown for clarity; in a real capacitor, the dielectric is in direct contact with the plates. Capacitors also allow AC current to flow and block DC current.

DC Sources

The dielectric between the plates is an insulator and blocks the flow of electrons. A steady current through a capacitor deposits electrons on one plate and removes the same quantity of electrons from the other plate. This process is commonly called 'charging' the capacitor. The current through the capacitor results in the separation of electric charge within the capacitor, which develops an electric field between the plates of the capacitor, equivalently, developing a voltage difference between the plates. This voltage V is directly proportional to the amount of charge separated Q. Since the current I through the capacitor is the rate at which charge Q is forced through the capacitor (dQ/dt), this can be expressed mathematically as:

$$I = \frac{dQ}{dt} = C\frac{dV}{dt} \quad \text{where}$$

I is the current flowing in the conventional direction, measured in amperes,

dV/dt is the time derivative of voltage, measured in volts per second, and

C is the capacitance in farads.

For circuits with a constant (DC) voltage source and consisting of only resistors and capacitors, the voltage across the capacitor cannot exceed the voltage of the source. Thus, an equilibrium is reached where the voltage across the capacitor is constant and the current through the capacitor is zero. For this reason, it is commonly said that capacitors block DC.

AC Sources

The current through a capacitor due to an AC source reverses direction periodically. That is, the alternating current alternately charges the plates: first in one direction and then the other. With the exception of the instant that the current changes direction, the capacitor current is non-zero at all times during a cycle. For this reason, it is commonly said that capacitors "pass" AC. However, at no time do electrons actually cross between the plates, unless the dielectric breaks down. Such a situation would involve physical damage to the capacitor and likely to the circuit involved as well.

Since the voltage across a capacitor is proportional to the integral of the current, as shown above, with sine waves in AC or signal circuits this results in a phase difference of 90 degrees, the current leading the voltage phase angle. It can be shown that the AC voltage across the capacitor is in quadrature with the alternating current through the capacitor. That is, the voltage and current are 'out-of-phase' by a quarter cycle. The amplitude of the voltage depends on the amplitude of the current divided by the product of the frequency of the current with the capacitance, C.

IMPEDANCE

The ratio of the phasor voltage across a circuit element to the phasor current through that element is called the impedance Z. For a capacitor, the impedance is given by

$$Z_C = \frac{V_C}{I_C} = \frac{-j}{2\pi f C} = -jX_C,$$

where

$X_C = \dfrac{1}{\omega C}$ is the *capacitive reactance,*

$\omega = 2\pi f$ is the angular frequency,

f is the frequency),

C is the capacitance in farads, and

j is the imaginary unit.

While this relation (between the *frequency domain* voltage and current associated with a capacitor) is always true, the ratio of the *time domain* voltage and current *amplitudes* is equal to X_C only for sinusoidal (AC) circuits in steady state.

Hence, capacitive reactance is the negative imaginary component of impedance. The negative sign indicates that the current leads the voltage by 90° for a sinusoidal signal, as opposed to the inductor, where the current lags the voltage by 90°.

The impedance is analogous to the resistance of a resistor. The impedance of a capacitor is inversely proportional to the frequency – that is, for very high-frequency alternating currents the reactance approaches zero – so that a capacitor is nearly a short circuit to a very high frequency AC source. Conversely, for very low frequency alternating currents, the reactance increases without bound so that a capacitor is nearly an open circuit to a very low frequency AC source. This frequency dependent behaviour accounts for most uses of the capacitor.

Reactance is so called because the capacitor doesn't dissipate power, but merely stores energy. In electrical circuits, as in mechanics, there are two types of load, resistive and reactive. Resistive loads (analogous to an object sliding on a rough surface) dissipate the energy delivered by the circuit as heat, while reactive loads (analogous to a spring or frictionless moving object) store this energy, ultimately delivering the energy back to the circuit.

Also significant is that the impedance is inversely proportional to the capacitance, unlike resistors and inductors for which impedances are linearly proportional to resistance and inductance respectively. This is why the series and shunt impedance formulae are the inverse of the resistive case. In series, impedances sum. In parallel, conductances sum.

LAPLACE EQUIVALENT

When using the Laplace transform in circuit analysis, the capacitive impedance is represented in the s domain by:

$$Z(s)=\frac{1}{Cs}$$

where C is the capacitance, and s ($=\sigma + j\omega$) is the complex frequency.

Displacement Current

The physicist James Clerk Maxwell invented the concept of displacement current, dD/dt, to make AmpΘre's law consistent with conservation of charge in cases where charge is accumulating as in a capacitor. He interpreted this as a real motion of charges, even in vacuum, where he supposed that it corresponded to motion of dipole charges in the aether. Although this interpretation has been abandoned, Maxwell's correction to AmpΘre's law remains valid.

NETWORKS

Series or Parallel Arrangements

Capacitors in a parallel configuration each have the same potential difference (voltage). Their total capacitance (C_{eq}) is given by:

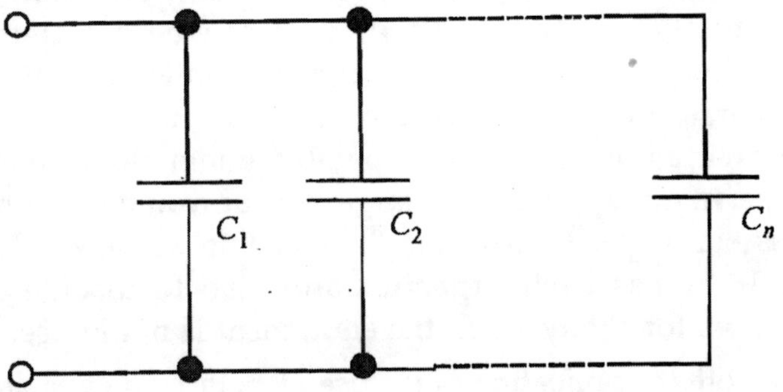

$$C_{eq} = C_1 + C_2 + ... + C_n$$

The reason for putting capacitors in parallel is to increase the total amount of charge stored. In other words, increasing the capacitance also increases the amount of energy that can be stored. Its expression is:

$$E_{\text{stored}} = \frac{1}{2}CV^2.$$

The current through capacitors in series stays the same, but the voltage across each capacitor can be different. The sum of the potential differences (voltage) is equal to the total voltage. Their total capacitance is given by:

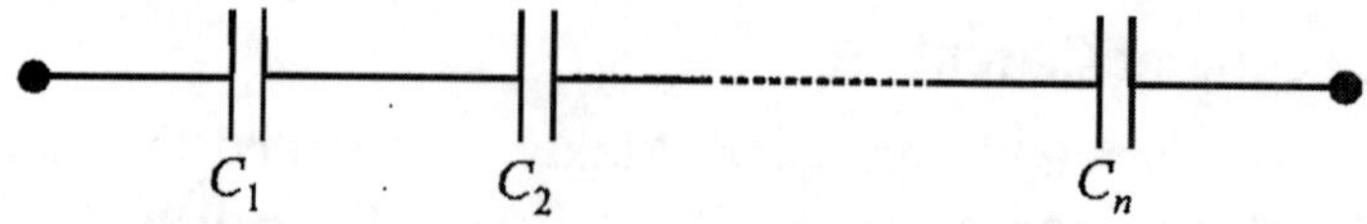

$$\frac{1}{C_{eq}} = \frac{1}{C_1} + \frac{1}{C_2} + \ldots + \frac{1}{C_n}$$

In parallel the effective area of the combined capacitor has increased, increasing the overall capacitance. While in series, the distance between the plates has effectively been increased, reducing the overall capacitance.

In practice capacitors will be placed in series as a means of economically obtaining very high voltage capacitors, for example for smoothing ripples in a high voltage power supply. Three "600 volt maximum" capacitors in series, will increase their overall working voltage to 1800 volts. This is of course offset by the capacitance obtained being only one third of the value of the capacitors used. This can be countered by connecting 3 of these series set-ups in parallel, resulting in a 3x3 matrix of capacitors with the same overall capacitance as an individual capacitor but operable under three times the voltage. In this application, a large resistor would be connected across each capacitor to ensure that the total voltage is divided equally across each capacitor and also to discharge the capacitors for safety when the equipment is not in use.

Another application is for use of polarized capacitors in alternating current circuits; the capacitors are connected in series, in reverse polarity, so that at any given time one of the capacitors is not conducting...

Capacitor/inductor Duality

In mathematical terms, the ideal capacitor can be considered as an inverse of the ideal inductor, because the voltage-current equations of the two devices can be transformed into one another by exchanging the voltage and current terms. Just as two or more inductors can be magnetically coupled to make a transformer, two or more charged conductors can be electrostatically coupled to make a capacitor. The *mutual capacitance* of two conductors is defined as the current that flows in one when the voltage across the other changes by unit voltage in unit time.

Capacitor Types

Fig. A 12 pF 20 kV fixed vacuum capacitor

Vacuum : Two metal, usually copper, electrodes are separated by a vacuum. The insulating envelope is usually glass or ceramic. Typically of low capacitance - 10 - 1000 pF and high voltage, up to tens of kilovolts, they are most often used in radio transmitters and other high voltage power devices. Both fixed and variable types are available. Vacuum variable capacitors can have a minimum to maximum capacitance ratio of up to 100, allowing any tuned circuit to cover a full decade of frequency. Vacuum is the most perfect of dielectrics with a zero loss tangent. This allows very high powers to be transmitted without significant loss and consequent heating.

- *Air:* Air dielectric capacitors consist of metal plates separated by an air gap. The metal plates, of which there may be many interleaved, are most often made of aluminium or silver-plated brass. Nearly all air dielectric capacitors are variable and are used in radio tuning circuits.
- *Metallized plastic film:* Made from high quality polymer film (usually polycarbonate, polystyrene, polypropylene, polyester (Mylar), and for high quality capacitors polysulfone), and metal foil or a layer of metal deposited on surface. They have good quality

and stability, and are suitable for timer circuits. Suitable for high frequencies.

- *Mica:* Similar to metal film. Often high voltage. Suitable for high frequencies. Expensive. Excellent tolerance.
- *Paper:* Used for relatively high voltages. Now obsolete.
- *Glass:* Used for high voltages. Expensive. Stable temperature coefficient in a wide range of temperatures.
- *Ceramic:* Chips of alternating layers of metal and ceramic. Depending on their dielectric,their degree of temperature/capacity dependence varies. They often have (especially the class 2) high dissipation factor, high frequency coefficient of dissipation, their capacity depends on applied voltage, and their capacity changes with aging. However they find massive use in common low-precision coupling and filtering applications. Suitable for high frequencies.
- *Aluminum electrolytic:* Polarized. Constructionally similar to metal film, but the electrodes are made of etched aluminium to acquire much larger surfaces. The dielectric is soaked with liquid electrolyte. They can achieve high capacities but suffer from poor tolerances, high instability, gradual loss of capacity especially when subjected to heat, and high leakage. Tend to lose capacity in low temperatures. Bad frequency characteristics make them unsuited for high-frequency applications. Special types with low equivalent series resistance are available.
- *Tantalum electrolytic:* Similar to the aluminum electrolytic capacitor but with better frequency and temperature characteristics. High dielectric absorption. High leakage. Has much better performance at low temperatures.
- *OS-CON* (or OS-CON) capacitors are a polymerized organic semiconductor solid-electrolyte type that offer longer life at higher cost than standard electrolytics.

- *Supercapacitors:* Made from carbon aerogel, carbon nanotubes, or highly porous electrode materials. Extremely high capacity. Can be used in some applications instead of rechargeable batteries.
- *Gimmick capacitors* are capacitors made from two insulated wires that have been twisted together. Each wire forms a capacitor plate. Gimmick capacitors are also a form of variable capacitor. Small changes in capacitance (20 percent or less) are obtained by twisting and untwisting the two wires.
- *Varicap* capacitors are specialized, reverse-biased diodes whose capacitance varies with voltage. Used in phase-locked loops, amongst other applications.
- varactor is actually a specialized reverse-biased diode used as a variable capacitor according to the voltage level applied.

Applications

- Capacitor symbols
- Capacitor Polarized Variable
- Capacitors capacitor

	capacitors	Capacitor
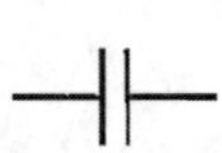	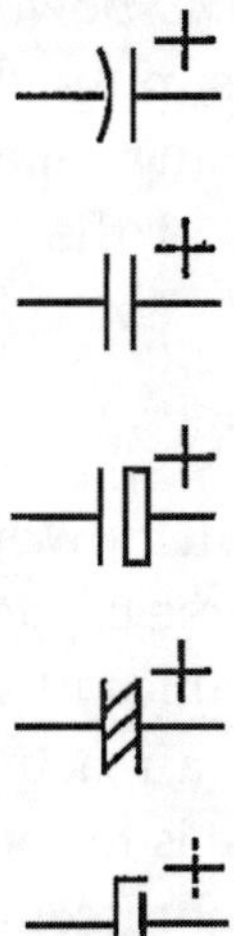	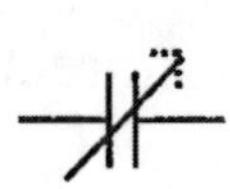

Capacitors have various uses in electronic and electrical systems.

Energy Storage

A capacitor can store electric energy when disconnected from its charging circuit, so it can be used like a temporary battery. Capacitors are commonly used in electronic devices to maintain power supply while batteries are being changed. (This prevents loss of information in volatile memory.)

Power Conditioning

Capacitors are used in power supplies where they smooth the output of a full or half wave rectifier. They can also be used in charge pump circuits as the energy storage element in the generation of higher voltages than the input voltage.

Capacitors are connected in parallel with the power circuits of most electronic devices and larger systems (such as factories) to shunt away and conceal current fluctuations from the primary power source to provide a "clean" power supply for signal or control circuits. Audio equipment, for example, uses several capacitors in this way, to shunt away power line hum before it gets into the signal circuitry. The capacitors act as a local reserve for the DC power source, and bypass AC currents from the power supply. This is used in car audio applications, when a stiffening capacitor compensates for the inductance and resistance of the leads to the lead-acid car battery.

Power factor correction

Capacitors are used in power factor correction. Such capacitors often come as three capacitors connected as a three phase load. Usually, the values of these capacitors are given not in farads but rather as a reactive power in volt-amperes reactive (VAr). The purpose is to counteract inductive loading from electric motors and fluorescent lighting in order to make the load appear to be mostly resistive.

FILTERING

Signal de-coupling

Because capacitors pass AC but block DC signals (when charged up to the applied dc voltage), they are often used to separate the AC and DC components of a signal. This method is known as *AC de-coupling*. Here, a large value of capacitance, whose value need not be accurately controlled, but whose reactance is small at the signal frequency, is employed.

Noise filters, motor starters, and snubbers

When an inductive circuit is opened, the current through the inductance collapses quickly, creating a large voltage across the open circuit of the switch or relay. If the inductance is large enough, the energy will generate a spark, causing the contact points to oxidize, deteriorate, or sometimes weld together, or destroying a solid-state switch. A snubber capacitor across the newly opened circuit creates a path for this impulse to bypass the contact points, thereby preserving their life; these were commonly found in contact breaker ignition systems, for instance. Similarly, in smaller scale circuits, the spark may not be enough to damage the switch but will still radiate undesirable radio frequency interference (RFI), which a filter capacitor absorbs. Snubber capacitors are usually employed with a low-value resistor in series, to dissipate energy and minimize RFI. Such resistor-capacitor combinations are available in a single package.

In an inverse fashion, to initiate current quickly through an inductive circuit requires a greater voltage than required to maintain it; in uses such as large motors, this can cause undesirable startup characteristics, and a motor starting capacitor is used to increase the coil current to help start the motor.

Capacitors are also used in parallel to interrupt units of a high-voltage circuit breaker in order to equally distribute the voltage between these units. In this case they are called grading capacitors.

In schematic diagrams, a capacitor used primarily for DC charge storage is often drawn vertically in circuit diagrams with the lower, more negative, plate drawn as an arc. The straight plate indicates the positive terminal of the device, if it is polarized.

Signal Processing

The energy stored in a capacitor can be used to represent information, either in binary form, as in DRAMs, or in analogue form, as in analog sampled filters and CCDs. Capacitors can be used in analog circuits as components of integrators or more complex filters and in negative feedback loop stabilization. Signal processing circuits also use capacitors to integrate a current signal.

Tuned circuits

Capacitors and inductors are applied together in tuned circuits to select information in particular frequency bands. For example, radio receivers rely on variable capacitors to tune the station frequency. Speakers use passive analog crossovers, and analog equalizers use capacitors to select different audio bands.

In a tuned circuit such as a radio receiver, the frequency selected is a function of the inductance (L) and the capacitance (C) in series, and is given by:

$$f = \frac{1}{2\pi\sqrt{LC}}$$

This is the frequency at which resonance occurs in an LC circuit.

OTHER APPLICATIONS

Sensing

Most capacitors are designed to maintain a fixed physical structure. However, various things can change the structure of the capacitor – the resulting change in capacitance can be used to sense those things.

Changing the dielectric: the effects of varying the physical and/or electrical characteristics of the dielectric can also be of use. Capacitors with an exposed and porous dielectric can be used to measure humidity in air.

Changing the distance between the plates: Capacitors are used to accurately measure the fuel level in airplanes. Capacitors with a flexible plate can be used to measure strain or pressure. Capacitors are used as the sensor in condenser microphones, where one plate is moved by air pressure, relative to the fixed position of the other plate. Some accelerometers use MEMS capacitors etched on a chip to measure the magnitude and direction of the acceleration vector. They are used to detect changes in acceleration, eg. as tilt sensors or to detect free fall, as sensors triggering airbag deployment, and in many other applications. Also some fingerprint sensors. Additionally, a user can adjust the pitch of a theremin musical instrument by moving his hand since this changes the effective capacitance between the users hand and the antenna.

Pulsed Power and Weapons

Groups of large, specially constructed, low-inductance high-voltage capacitors (*capacitor banks*) are used to supply huge pulses of current for many pulsed power applications. These include electromagnetic forming, Marx generators, pulsed lasers (especially TEA lasers), pulse forming networks, radar, fusion research, and particle accelerators.

Large capacitor banks(Reservoir) are used as energy sources for the exploding-bridgewire detonators or slapper detonators in nuclear weapons and other specialty weapons. Experimental work is under way using banks of capacitors as power sources for electromagnetic armour and electromagnetic railguns or coilguns.

Hazards and Safety

Capacitors may retain a charge long after power is removed from a circuit; this charge can cause shocks (sometimes fatal) or damage to connected equipment. For example, even a seemingly innocuous device such as a disposable camera flash unit powered by a 1.5 volt AA battery contains a capacitor which may be charged to over 300 volts. This is easily capable of delivering an extremely painful, and possibly lethal shock.

Care must be taken to ensure that any large or high-voltage capacitor is properly discharged before servicing the containing equipment. For board-level capacitors, this is done by placing a bleeder resistor across the terminals, whose resistance is large enough that the leakage current will not affect the circuit, but small enough to discharge the capacitor shortly after power is removed. High-voltage capacitors should be stored with the terminals shorted, since temporarily discharged capacitors can develop potentially dangerous voltages when the terminals are left open-circuited.

Large oil-filled old capacitors must be disposed of properly as some contain polychlorinated biphenyls (PCBs). It is known that waste PCBs can leak into groundwater under landfills. If consumed by drinking contaminated water, PCBs are carcinogenic, even in very tiny amounts. If the capacitor is physically large it is more likely to be dangerous and may require precautions in addition to those described above. New electrical components are no longer produced with PCBs. ("PCB" in electronics usually means printed circuit board, but the above usage is an exception.) Capacitors containing PCB were labelled as containing "Askarel" and several other trade names.

High-voltage

Above and beyond usual hazards associated with working with high voltage, high energy circuits, there are a number of dangers that are specific to high voltage capacitors. High voltage capacitors may catastrophically fail when subjected to voltages or currents beyond their rating, or as they reach their normal end of life. Dielectric or metal interconnection failures may create arcing within oil-filled units that vapourizes dielectric fluid, resulting in case bulging, rupture, or even an explosion that disperses flammable oil, starts fires, and damages nearby equipment. Rigid cased cylindrical glass or plastic cases are more prone to explosive rupture than rectangular cases due to an inability to easily expand under pressure. Capacitors used in RF or sustained high current applications can overheat, especially in the centre of the capacitor rolls. The trapped heat may cause rapid interior heating and destruction, even though the outer case remains relatively cool. Capacitors used within high energy capacitor banks can violently explode when a fault in one capacitor causes sudden dumping of energy stored in the rest of the bank into the failing unit. And, high voltage vacuum capacitors can generate soft X-rays even during normal operation. Proper containment, fusing, and preventative maintenance can help to minimize these hazards.

High voltage capacitors can benefit from a pre-charge to limit in-rush currents at power-up of HVDC circuits. This will extend the life of the component and may mitigate high voltage hazards.

DYADICS

A dyadic, also known as a vector direct product, is a linear polynomial of dyads $AB + CD + \ldots$ consisting of nine components A_{ij} which transform as

$$(A_{ij})' = \sum_{m,n} \frac{h_m h_n}{h_i' h_j'} \frac{\partial x_m}{\partial x_i'} \frac{\partial x_n}{\partial x_j'} A_{mn}$$

$$= \sum_{m,n} \frac{h_i' h_j'}{h_m h_n} \frac{\partial x_i'}{\partial x_m} \frac{\partial x_j'}{\partial x_n} A_{mn}$$

$$= \sum_{m,n} \frac{h_i' h_n}{h_m h_j'} \frac{\partial x_i'}{\partial x_m} \frac{\partial x_n}{\partial x_j'} A_{mn}.$$

Dyadics are often represented by Gothic capital letters. The use of dyadics is nearly archaic since tensors perform the same function but are notationally simpler.

A unit dyadic is also called the idemfactor and is defined such that

$$I.A \equiv A$$

In Cartesian coordinates,

$$I = \hat{x}\hat{x} + \hat{y}\hat{y} + \hat{z}\hat{z},$$

and in spherical coordinates

$$I = \nabla r.$$

Dyadic Product

In mathematics, in particular multilinear algebra, the dyadic product

$$P = u \otimes v$$

of two vectors, u and v, each having the same dimension, is the tensor product of the vectors and results in a tensor of rank two.

COMPONENTS

With respect to a chosen basis $\{e_i\}$, the components P_{ij} of the dyadic product $P = u \otimes v$ may be defined by

$$P_{ij} = u_i v_j,$$

where

$$u = \sum_i u_i e_i,$$

$$v = \sum_j v_j e_j,$$

and

$$P = \sum_{ij} P_{ij} e_i \otimes e_j.$$

Matrix Representation

The dyadic product can be simply represented as the square matrix obtained by multiplying u as a column vector by v as a row vector. For example,

$$u \otimes v \rightarrow \begin{bmatrix} u_1 \\ u_2 \\ u_3 \end{bmatrix} \begin{bmatrix} v_1 & v_2 & v_3 \end{bmatrix} = \begin{bmatrix} u_1v_1 & u_1v_2 & u_1v_3 \\ u_2v_1 & u_2v_2 & u_2v_3 \\ u_3v_1 & u_3v_2 & u_3v_3 \end{bmatrix},$$

where the arrow indicates that this is only one particular representation of the dyadic product, referring to a particular basis. In this representation, the dyadic product is a special case of the Kronecker product.

Dyadic Rational

In mathematics, a dyadic fraction or dyadic rational is a rational number whose denominator is a power of two, i.e., a number of the form $a/2^b$ where a is an integer and b is a natural number; for example, 1/2 or 3/8, but not 1/3. These are precisely the numbers whose binary expansion is finite.

The inch is customarily subdivided in dyadic rather than decimal fractions; similarly, the customary divisions of the gallon into half-gallons, quarts, and pints are dyadic. The ancient Egyptians also used dyadic fractions in measurement, with denominators up to 1/64, using a notation based on the Eye of Horus.

The set of all dyadic fractions is dense in the real line: any real number x can be arbitrarily closely approximated by dyadic rationals of the form $\lfloor 2^i x \rfloor / 2^i$. Compared to other dense subsets of the real line, such as the rational numbers, the dyadic rationals are in some sense a relatively "small" dense set, which is why they sometimes occur in proofs. (See for instance Urysohn's lemma.)

The sum, product, or difference of any two dyadic fractions is itself another dyadic fraction:

$$\frac{a}{2^b} + \frac{c}{2^d} = \frac{2^{d-b}a + c}{2^d} \quad (d \geq b)$$

$$\frac{a}{2^b} - \frac{c}{2^d} = \frac{2^{d-b}a - c}{2^d} \quad (d \geq b)$$

$$\frac{a}{2^b} - \frac{c}{2^d} = \frac{a - 2^{b-d}c}{2^b} \quad (d < b)$$

$$\frac{a}{2^b} \times \frac{c}{2^d} = \frac{a \times c}{2^{b+d}}$$

However, the result of dividing one dyadic fraction by another is, in general, not a dyadic fraction. Thus, the dyadic fractions form a subring of the rational numbers Q. Algebraically, this subring is the localization of the integers Z with respect to the set of powers of two.

The surreal numbers are generated by an iterated construction principle which starts by generating all finite dyadic fractions, and then goes on to create new and strange kinds of infinite, infinitesimal and other numbers.

Dyadic Solenoid

As an additive abelian group the dyadic rationals are the direct limit of infinite cyclic subgroups of the rational numbers,

$$\varinjlim \{2^{-i}\mathbb{Z} \mid i = 0,1,2,...\}$$

In the spirit of Pontryagin duality, there is a dual object, namely the inverse limit of the unit circle group under the repeated squaring map

$$\varsigma \to \varsigma^2.$$

The resulting topological group D is called the dyadic solenoid.

An element of the dyadic solenoid can be represented as an infinite sequence of complex numbers $q_0, q_1, q_2, ...$, with the properties that each q_i lies on the unit circle and that, for all $i > 0$, $q_i^2 = q_{i-1}$. The group operation on these elements multiplies any two sequences componentwise.

As a topological space the dyadic solenoid is an indecomposable continuum.

5

Special Relativity

The special theory of relativity was proposed in 1905 by Albert Einstein in his article "On the Electrodynamics of Moving Bodies". Some three centuries earlier, Galileo's principle of relativity had stated that all uniform motion was relative, and that there was no absolute and well-defined state of rest; a person on the deck of a ship may be at rest in his opinion, but someone observing from the shore would say that he was moving. Einstein's theory generalized Galilean relativity from only mechanics to all laws of physics including electrodynamics. To stress this point, Einstein not only widened the postulate of relativity, but added the second postulate - that all observers will always measure the speed of light to be the same no matter what their state of uniform linear motion.

This theory has a variety of surprising consequences that seem to violate common sense, but all have been experimentally verified. Special relativity overthrows Newtonian notions of absolute space and time by stating that distance and time depend on the observer, and that time and space are perceived differently, depending on the observer. It yields the equivalence of matter and energy, as expressed in the mass-energy equivalence formula $E = mc^2$, where c is the speed of light in a vacuum. Special relativity agrees with Newtonian mechanics in their common realm of applicability, in experiments in which all velocities are small compared to the speed of light.

The theory was called "special" because it applies the principle of relativity only to inertial frames. Einstein developed general relativity to apply the principle generally, that is, to any frame, and that theory includes the effects of gravity. Special relativity does not account for gravity, but it can deal with accelerations.

Although special relativity makes some quantities relative, such as time, that we would have imagined to be absolute based on everyday experience, it also makes absolute some others that we would have thought were relative. In particular, it states that the speed of light is the same for all observers, even if they are in motion relative to one another. Special relativity reveals that c is not just the velocity of a certain phenomenon - light - but rather a fundamental feature of the way space and time are tied together. In particular, special relativity states that it is impossible for any material object to accelerate to light speed.

Postulates

- First postulate - Special principle of relativity - The laws of physics are the same in all inertial frames of reference. In other words, there are no privileged inertial frames of reference.
- Second postulate - Invariance of c - The speed of light in a vacuum is a universal constant, c, which is independent of the motion of the light source.

The power of Einstein's argument stems from the manner in which he derived startling and seemingly implausible results from two simple assumptions that were founded on analysis of observations. An observer attempting to measure the speed of light's propagation will get exactly the same answer no matter how the observer or the system's components are moving.

Lack of an Absolute Reference Frame

The principle of relativity, which states that there is no stationary reference frame, dates back to Galileo, and was incorporated into Newtonian Physics. However, in the late

19th century, the existence of electromagnetic waves led some physicists to suggest that the universe was filled with a substance known as "aether", which would act as the medium through which these waves, or vibrations traveled. The aether was thought to constitute an absolute reference frame against which speeds could be measured. In other words, the aether was the only fixed or motionless thing in the universe. Aether supposedly had some wonderful properties: it was sufficiently elastic that it could support electromagnetic waves, and those waves could interact with matter, yet it offered no resistance to bodies passing through it.

The results of various experiments, including the Michelson-Morley experiment, indicated that the Earth was always 'stationary' relative to the aether – something that was difficult to explain, since the Earth is in orbit around the Sun. Einstein's elegant solution was to discard the notion of an aether and an absolute state of rest. Special relativity is formulated so as to not assume that any particular frame of reference is special; rather, in relativity, any reference frame moving with uniform motion will observe the same laws of physics. In particular, the speed of light in a vacuum is always measured to be c, even when measured by multiple systems that are moving at different (but constant) velocities.

Consequences

Einstein has said that all of the consequences of special relativity can be derived from examination of the Lorentz transformations.

These transformations, and hence special relativity, lead to different physical predictions than Newtonian mechanics when relative velocities become comparable to the speed of light. The speed of light is so much larger than anything humans encounter that some of the effects predicted by relativity are initially counter-intuitive:

- *Time dilation*: The time lapse between two events is not invariant from one observer to another, but is dependent on the relative speeds of the observers' reference frames (e.g., the twin paradox which concerns a twin who flies off in a spaceship traveling near the speed of light and returns to discover that his twin has aged much more).

- *Relativity of simultaneity*: Two events happening in two different locations that occur simultaneously to one observer, may occur at different times to another observer (lack of absolute simultaneity).
- *Lorentz contraction*: The dimensions (e.g., length) of an object as measured by one observer may be smaller than the results of measurements of the same object made by another observer (e.g., the ladder paradox involves a long ladder traveling near the speed of light and being contained within a smaller garage).
- ***Composition of velocities:*** Velocities (and speeds) do not simply 'add', for example if a rocket is moving at T! the speed of light relative to an observer, and the rocket fires a missile at T! of the speed of light relative to the rocket, the missile does not exceed the speed of light relative to the observer. (In this example, the observer would see the missile travel with a speed of 12/13 the speed of light.)
- *Inertia and momentum*: As an object's velocity approaches the speed of light from an observer's point of view, its mass appears to increase thereby making it more and more difficult to accelerate it from within the observer's frame of reference.
- *Equivalence of mass and energy, $E = mc^2$* – The energy content of an object at rest with mass m equals mc^2. Conservation of energy implies that in any reaction a decrease of the sum of the masses of particles must be accompanied by an increase in kinetic energies of the particles after the reaction. Similarly, the mass of an object can be increased by taking in kinetic energies.

REFERENCE FRAMES, COORDINATES AND THE LORENTZ TRANSFORMATION

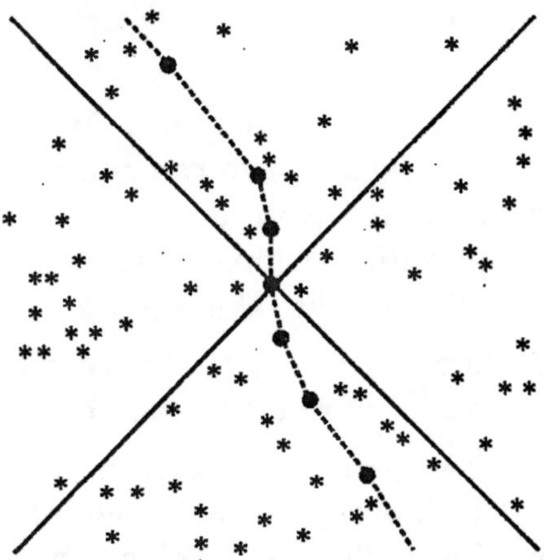

Fig. Changing views of spacetime along the world line of a rapidly accelerating observer.

In this animation, the vertical direction indicates time and the horizontal direction indicates distance, the dashed line is the spacetime trajectory ("world line") of the observer. The lower quarter of the diagram shows the events that are visible to the observer, and the upper quarter shows the light cone-those that will be able to see the observer. The small dots are arbitrary events in spacetime.

The slope of the world line (deviation from being vertical) gives the relative velocity to the observer. Note how the view of spacetime changes when the observer accelerates.

Relativity theory depends on "reference frames". A reference frame is an observational perspective in space at rest, or in uniform motion, from which a position can be measured along 3 spatial axes. In addition, a reference frame has the ability to determine measurements of the time of events using a 'clock' (any reference device with uniform periodicity).

An event is an occurrence that can be assigned a single unique time and location in space relative to a reference frame: it is a "point" in space-time. Since the speed of light is constant in relativity in each and every reference frame, pulses of light can be used to unambiguously measure distances and refer back the times that events occurred to the clock, even though light takes time to reach the clock after the event has transpired.

For example, the explosion of a firecracker may be considered to be an "event". We can completely specify an event by its four space-time coordinates: The time of occurrence and its 3-dimensional spatial location define a reference point. Let's call this reference frame S.

In relativity theory we often want to calculate the position of a point from a different reference point.

Suppose we have a second reference frame S′, whose spatial axes and clock exactly coincide with that of S at time zero, but it is moving at a constant velocity v with respect to S along the x-axis.

Since there is no absolute reference frame in relativity theory, a concept of 'moving' doesn't strictly exist, as everything is always moving with respect to some other reference frame. Instead, any two frames that move at the same speed in the same direction are said to be *comoving*. Therefore S and S′ are not *comoving*.

Let's define the event to have space-time coordinates (t, x, y, z) in system S and (t', x', y', z') in S'. Then the Lorentz transformation specifies that these coordinates are related in the following way:

$$t' = \gamma\left(t - \frac{vx}{c^2}\right)$$

$$x' = \gamma(x - vt)$$

$$y' = y$$

$$z' = z$$

where $\gamma = \frac{1}{\sqrt{1 - v^2/c^2}}$ is called the Lorentz factor and is the speed of light in a vacuum.

The y and z coordinates are unaffected, but the x and t axes are mixed up by the transformation. In a way this transformation can be understood as a hyperbolic rotation.

A quantity invariant under Lorentz transformations is known as a Lorentz scalar.

Simultaneity

From the first equation of the Lorentz transformation in terms of coordinate differences

$$\Delta t' = \gamma\left(\Delta t - \frac{v\Delta x}{c^2}\right)$$

it is clear that two events that are simultaneous in frame S (satisfying $\Delta t = 0$), are not necessarily simultaneous in another inertial frame S' (satisfying $\Delta t' = 0$). Only if these events are colocal in frame S (satisfying $\Delta x = 0$), will they be simultaneous in another frame S'.

Time Dilation and Length Contraction

Writing the Lorentz Transformation and its inverse in terms of coordinate differences we get

$$\Delta t' = \gamma\left(\Delta t - \frac{v\Delta x}{c^2}\right)$$

$$\Delta x' = \gamma(\Delta x - v\Delta t)$$

and

$$\Delta t = \gamma\left(\Delta t' + \frac{v\Delta x'}{c^2}\right)$$

$$\Delta x' = \gamma(\Delta x' + v\Delta t')$$

Suppose we have a clock at rest in the unprimed system S. Two consecutive ticks of this clock are then characterized by $\Delta x = 0$. If we want to know the relation between the times between these ticks as measured in both systems, we can use the first equation and find:

$$\Delta t' = \gamma\Delta t \qquad \text{(for events satisfying } \Delta x = 0)$$

This shows that the time $\Delta t'$ between the two ticks as seen in the 'moving' frame S' is larger than the time Δt between these ticks as measured in the rest frame of the clock. This phenomenon is called time dilation.

Similarly, suppose we have a measuring rod at rest in the unprimed system. In this system, the length of this rod is written as Δx. If we want to find the length of this rod as measured in the 'moving' system S', we must make sure to measure the distances x' to the end points of the rod simultaneously in the primed frame S'. In other words, the measurement is characterized by $\Delta t' = 0$, which we can combine with the fourth equation to find the relation between the lengths Δx and $\Delta x'$:

$$\Delta x' = \frac{\Delta x}{\gamma} \qquad \text{(for events satisfying } \Delta t' = 0\text{)}$$

This shows that the length $\Delta x'$ of the rod as measured in the 'moving' frame S' is shorter than the length Δx in its own rest frame. This phenomenon is called *length contraction* or *Lorentz contraction.*

These effects are not merely appearances; they are explicitly related to our way of measuring *time intervals* between events which occur at the same place in a given coordinate system (called "co-local" events). These time intervals will be *different* in another coordinate system moving with respect to the first, unless the events are also simultaneous. Similarly, these effects also relate to our measured distances between separated but simultaneous events in a given coordinate system of choice. If these events are not co-local, but are separated by distance (space), they will *not* occur at the same *spacial distance* from each other when seen from another moving coordinate system.

CAUSALITY AND PROHIBITION OF MOTION FASTER THAN LIGHT

In the fig. the interval AB is 'time-like'; *i.e.*, there is a frame of reference in which event A and event B occur at the same location in space, separated only by occurring at different times. If A precedes B in that frame, then A precedes B in all frames. It is hypothetically possible for matter (or information) to travel from A to B, so there can be a causal relationship (with A the cause and B the effect).

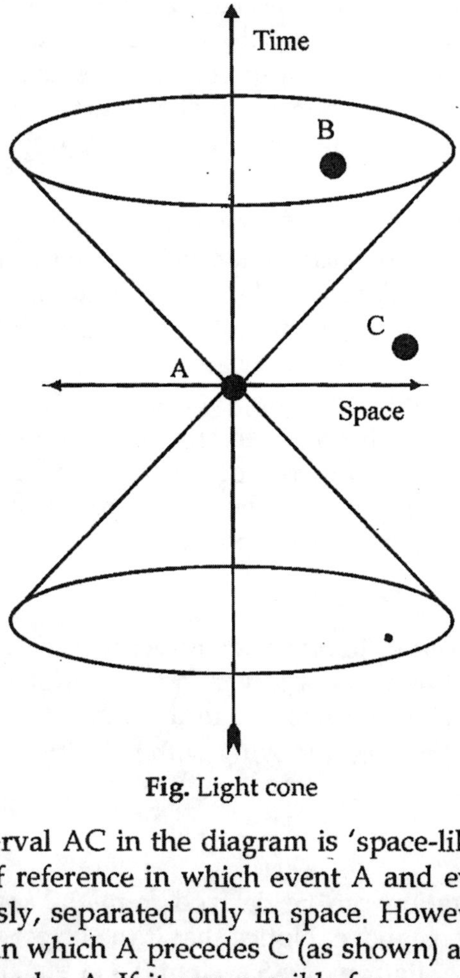

Fig. Light cone

The interval AC in the diagram is 'space-like'; *i.e.*, there is a frame of reference in which event A and event C occur simultaneously, separated only in space. However there are also frames in which A precedes C (as shown) and frames in which C precedes A. If it was possible for a cause-and-effect relationship to exist between events A and C, then paradoxes of causality would result. For example, if A was the cause, and C the effect, then there would be frames of reference in which the effect preceded the cause. Although this in itself won't give rise to a paradox, one can show that faster than light signals can be sent back into one's own past. A causal paradox can then be constructed by sending the signal if and only if no signal was received previously.

Therefore, one of the consequences of special relativity is that (assuming causality is to be preserved), no information or material object can travel faster than light. On the other hand, the logical situation is not as clear in the case of general relativity, so it is an open question whether or not there is some fundamental principle that preserves causality (and therefore prevents motion faster than light) in general relativity.

Even without considerations of causality, there are other strong reasons why faster-than-light travel is forbidden by special relativity. For example, if a constant force is applied to an object for a limitless amount of time, then integrating $F = dp/dt$ gives a momentum that grows without bound, but this is simply because $p = m\gamma v$ approaches infinity as v approaches c. To an observer who is not accelerating, it appears as though the object's inertia is increasing, so as to produce a smaller acceleration in response to the same force. This behaviour is in fact observed in particle accelerators.

Composition of Velocities

If the observer in S sees an object moving along the axis at velocity ω, then the observer in the S' system, a frame of reference moving at velocity v in the x direction with respect to S, will see the object moving with velocity ω' where

$$\omega' = \frac{\omega - v}{1 - \omega v / c^2}.$$

This equation can be derived from the space and time transformations above. Notice that if the object were moving at the speed of light in the S system (i.e. $\omega = c$), then it would also be moving at the speed of light in the S' system. Also, if both and are small with respect to the speed of light, we will recover the intuitive Galilean transformation of velocities: $\omega' \approx \omega - v$.

Mass, Momentum, and Energy

In addition to modifying notions of space and time, special relativity forces one to reconsider the concepts of mass, momentum, and energy, all of which are important constructs in Newtonian mechanics. Special relativity shows, in fact, that these concepts are all different aspects of the same physical quantity in much the same way that it shows space and time to be interrelated.

There are a couple of (equivalent) ways to define momentum and energy in SR. One method uses conservation laws. If these laws are to remain valid in SR they must be true in every possible reference frame. However, if one does some simple thought experiments using the Newtonian definitions of momentum and energy one sees that these quantities are not conserved in SR. One can rescue the idea of conservation by making some small modifications to the definitions to account for relativistic velocities. It is these new definitions which are taken as the correct ones for momentum and energy in SR.

Given an object of invariant mass m traveling at velocity v the energy and momentum are given (and even defined) by

$$E = \gamma mc^2$$

$$\vec{p} = \gamma m\vec{v}$$

where γ (the Lorentz factor) is given by

$$\gamma = \frac{1}{\sqrt{1-\beta^2}}$$

where β is the velocity as a ratio of the speed of light. The term γ occurs frequently in relativity, and comes from the Lorentz transformation equations.

Relativistic energy and momentum can be related through the formula

$$E^2 - (pc)^2 = (mc^2)^2$$

which is referred to as the *relativistic energy-momentum equation.* It is interesting to observe that while the energy E and the momentum p are observer dependent (vary from frame to frame) the quantity $E^2 - (pc)^2 = (mc^2)^2$ is observer independent.

For velocities much smaller than those of light, γ can be approximated using a Taylor series expansion and one finds that

$$E \approx mc^2 + \frac{1}{2}mv^2$$

- $\vec{p} \approx m\vec{v}$

Barring the first term in the energy expression, these formulas agree exactly with the standard definitions of Newtonian kinetic energy and momentum. This is as it should be, for special relativity must agree with Newtonian mechanics at low velocities.

Looking at the above formulas for energy, one sees that when an object is at rest ($v = 0$ and $\gamma = 1$) there is a non-zero energy remaining:

$$E_{rest} = mc^2$$

This energy is referred to as *rest energy*. The rest energy does not cause any conflict with the Newtonian theory because it is a constant and, as far as kinetic energy is concerned, it is only differences in energy which are meaningful.

Taking this formula at face value, we see that in relativity, *mass is simply another form of energy*. In 1927 Einstein remarked about special relativity:

Under this theory mass is not an unalterable magnitude, but a magnitude dependent on (and, indeed, identical with) the amount of energy.

This formula becomes important when one measures the masses of different atomic nuclei. By looking at the difference in masses, one can predict which nuclei have extra stored energy that can be released by nuclear reactions, providing important information which was useful in the development of nuclear energy and, consequently, the nuclear bomb. The implications of this formula on 20th-century life have made it one of the most famous equations in all of science.

Relativistic Mass

Introductory physics courses and some older textbooks on special relativity sometimes define a *relativistic mass* which increases as the velocity of a body increases. According to the geometric interpretation of special relativity, this is often deprecated and the term 'mass' is reserved to mean invariant mass and is thus independent of the inertial frame, i.e., *invariant*.

Using the relativistic mass definition, the mass of an object may vary depending on the observer's inertial frame in the same way that other properties such as its length may do so. Defining such a quantity may sometimes be *useful* in that doing so simplifies a calculation by restricting it to a specific frame. For example, consider a body with an invariant mass m moving at some velocity relative to an observer's reference system. That observer defines the *relativistic mass* of that body as:

$$M = \gamma m$$

"Relativistic mass" should not be confused with the "longitudinal" and "transverse mass" definitions that were used around 1900 and that were based on an inconsistent application of the laws of Newton: those used $f = ma$ for a variable mass, while relativistic mass corresponds to Newton's dynamic mass in which $p = Mv$ and $f = dp/dt$.

Note also that the body does *not* actually become more massive in its *proper* frame, since the relativistic mass is only different for an observer in a different frame. The *only* mass that is frame independent is the invariant mass. When using the relativistic mass, the applicable reference frame should be specified if it isn't already obvious or implied. It also goes almost without saying that the increase in relativistic mass does not come from an increased number of atoms in the object. Instead, the relativistic mass of each atom and subatomic particle has increased.

Physics textbooks sometimes use the relativistic mass as it allows the students to utilize their knowledge of Newtonian physics to gain some intuitive grasp of relativity in their frame of choice (usually their own!). "Relativistic mass" is also consistent with the concepts "time dilation" and "length contraction".

FORCE

The classical definition of ordinary force f is given by Newton's Second Law in its original form:

$$\vec{f} = d\,\vec{p}\,/\,dt$$

and this is valid in relativity.

Many modern textbooks rewrite Newton's Second Law as

$$\vec{f} = M\vec{a}$$

This form is not valid in relativity or in other situations where the relativistic mass M is varying.

This formula can be replaced in the relativistic case by

$$\vec{f} = \gamma m\vec{a} + \gamma^3 m\frac{\vec{v}.\vec{a}}{c^2}\vec{v}$$

As seen from the equation, ordinary force and acceleration vectors are not necessarily parallel in relativity.

However the four-vector expression relating four-force F^{μ} to invariant mass m and four-acceleration A^{μ} restores the same equation form

$$F^{\mu} = m\,A^{\mu}$$

THE GEOMETRY OF SPACE-TIME

SR uses a 'flat' 4-dimensional Minkowski space, which is an example of a space-time. This space, however, is very similar to the standard 3 dimensional Euclidean space, and fortunately by that fact, very easy to work with.

The differential of distance (ds) in cartesian 3D space is defined as:

$$ds^2 = dx_1^2 + dx_2^2 + dx_3^2$$

where (dx_1, dx_2, dx_3) are the differentials of the three spatial dimensions. In the geometry of special relativity, a fourth dimension is added, derived from time, so that the equation for the differential of distance becomes:

$$ds^2 = dx_1^2 + dx_2^2 + dx_3^2 - c^2dt^2$$

If we wished to make the time coordinate look like the space coordinates, we could treat time as imaginary: $x_4 = ict$. In this case the above equation becomes symmetric:

$$ds^2 = dx_1^2 + dx_2^2 + dx_3^2 + dx_4^2$$

This suggests what is in fact a profound theoretical insight as it shows that special relativity is simply a rotational symmetry of our space-time, very similar to rotational symmetry of Euclidean space. Just as Euclidean space uses a Euclidean metric, so space-time uses a Minkowski metric. Basically, SR can be stated in terms of the invariance of space-time interval (between any two events) as seen from any inertial reference frame. All equations and effects of special relativity can be derived from this rotational symmetry (the Poincaré group) of Minkowki space-time. According to Misner (1971 §2.3), ultimately the deeper understanding of both special and general relativity will come from the study of the Minkowski metric (described below) rather than a "disguised" Euclidean metric using *ict* as the time coordinate.

If we reduce the spatial dimensions to 2, so that we can represent the physics in a 3-D space

$$ds^2 = dx_1^2 + dx_2^2 - c^2dt^2$$

We see that the null geodesics lie along a dual-cone:

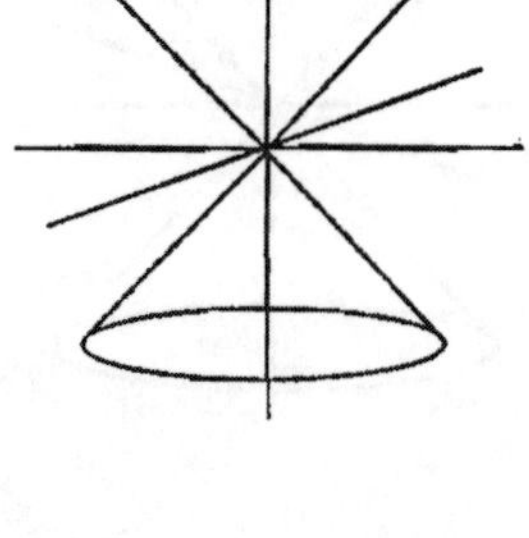

defined by the equation

$$ds^2 = 0 = dx_1^2 + dx_2^2 - c^2 dt^2$$

or

$$dx_1^2 + dx_2^2 = c^2 dt^2$$

Which is the equation of a circle with $r = c \times dt$. If we extend this to three spatial dimensions, the null geodesics are the 4-dimensional cone:

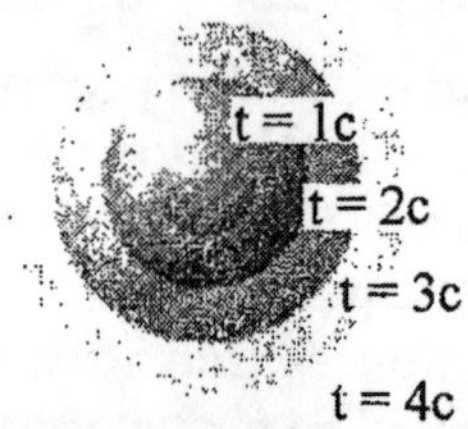

$$dx_1^2 + dx_2^2 + dx_3^2 = c^2 dt^2$$

This null dual-cone represents the "line of sight" of a point in space. That is, when we look at the stars and say "The light from that star which I am receiving is X years old", we are looking down this line of sight: a null geodesic. We are looking at an event $d = \sqrt{x_1^2 + x_2^2 + x_3^2}$ meters away and d/c seconds in the past. For this reason the null dual cone is also known as the 'light cone'. (The point in the lower left of the picture below represents the star, the origin represents the observer, and the line represents the null geodesic "line of sight".)

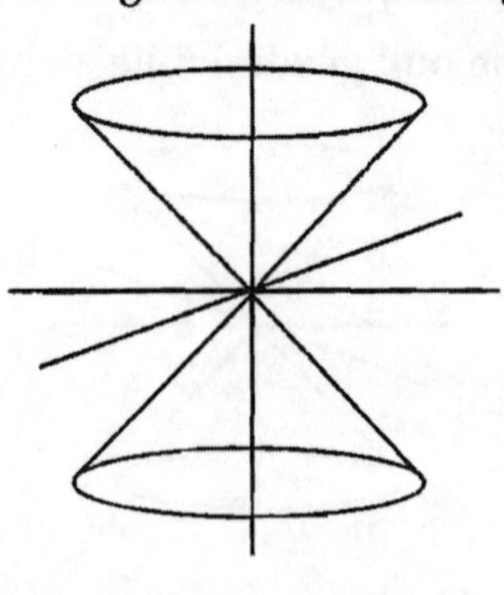

The cone in the $-t$ region is the information that the point is 'receiving', while the cone in the $+t$ section is the information that the point is 'sending'.

The geometry of Minkowski space can be depicted using Minkowski diagrams, which are also useful in understanding many of the thought-experiments in special relativity.

PHYSICS IN SPACETIME

Here, we see how to write the equations of special relativity in a manifestly Lorentz covariant form. The position of an event in spacetime is given by a contravariant four vector whose components are:

$$x^v = (t, x, y, z)$$

That is, $x^0 = t$ and $x^1 = x$ and $x^2 = y$ and $x^3 = z$. Superscripts are contravariant indices in this section rather than exponents except when they indicate a square. Subscripts are covariant indices which also range from zero to three as with the spacetime gradient of a field φ :

$$\partial_0\phi = \frac{\partial\phi}{\partial t},\ \partial_1\phi = \frac{\partial\phi}{\partial x},\ \partial_2\phi = \frac{\partial\phi}{\partial y},\ \partial_3\phi = \frac{\partial\phi}{\partial z}.$$

Metric and Transformations of Coordinates

Having recognised the four-dimensional nature of spacetime, we are driven to employ the Minkowski metric, η, given in components (valid in any inertial reference frame) as:

$$\eta\alpha\beta = \begin{pmatrix} -c^2 & 0 & 0 & 0 \\ 0 & 1 & 0 & 0 \\ 0 & 0 & 1 & 0 \\ 0 & 0 & 0 & 1 \end{pmatrix}$$

Its reciprocal is:

$$\eta^{\alpha\beta} = \begin{pmatrix} -1/c^2 & 0 & 0 & 0 \\ 0 & 1 & 0 & 0 \\ 0 & 0 & 1 & 0 \\ 0 & 0 & 0 & 1 \end{pmatrix}$$

Then we recognize that co-ordinate transformations between inertial reference frames are given by the Lorentz transformation tensor Λ. For the special case of motion along the x-axis, we have:

$$\Lambda^{\mu'}{}_{v} = \begin{pmatrix} \gamma & -\beta\gamma/c & 0 & 0 \\ -\beta\gamma c & \gamma & 0 & 0 \\ 0 & 0 & 1 & 0 \\ 0 & 0 & 0 & 1 \end{pmatrix}$$

which is simply the matrix of a boost (like a rotation) between the x and t coordinates. Where μ' indicates the row and v indicates the column. Also, β and γ are defined as:

$$\beta = \frac{v}{c}, \gamma = \frac{1}{\sqrt{1-\beta^2}}.$$

More generally, a transformation from one inertial frame (ignoring translations for simplicity) to another must satisfy:

$$\eta_{\alpha\beta} = \eta_{\mu' v'} \Lambda^{\mu'}{}_{\alpha} \Lambda^{v'}{}_{\beta}$$

where there is an implied summation of μ' and v' from 0 to 3 on the right-hand side in accordance with the Einstein summation convention. The Poincaré group is the most general group of transformations which preserves the Minkowski metric and this is the physical symmetry underlying special relativity.

All proper physical quantities are given by tensors. So to transform from one frame to another, we use the well known tensor transformation law

$$T^{[i'_1,i'_2,\ldots i'_p]}_{[j'_1,j'_2,\ldots j'_q]} = \Lambda^{i'_1}{}_{i_1}\Lambda^{i'_2}{}_{i_2}\ldots\Lambda^{i'_p}{}_{i_p}\Lambda_{j'_1}{}^{j_1}\Lambda_{j'_2}{}^{j_2}$$

$$\ldots\Lambda_{j'_q}{}^{j_q}\, T^{[i_1,i_2\ldots i_p]}_{[j_1,j_2\ldots j_q]}$$

Where $\Lambda_{j'_k}{}^{j_k}$ is the reciprocal matrix of $\Lambda^{j_k}{}_{j_k}$.

To see how this is useful, we transform the position of an event from an unprimed co-ordinate system S to a primed system S', we calculate

$$\begin{pmatrix} t' \\ x' \\ y' \\ z' \end{pmatrix} = x^{\mu'} = \Lambda^{\mu'}{}_{\nu} x^{\nu} = \begin{pmatrix} \gamma & -\beta\gamma/c & 0 & 0 \\ -\beta\gamma c & \gamma & 0 & 0 \\ 0 & 0 & 1 & 0 \\ 0 & 0 & 0 & 1 \end{pmatrix} \begin{pmatrix} t \\ x \\ y \\ z \end{pmatrix}$$

$$= \begin{pmatrix} \gamma t - \gamma\beta x/c \\ \gamma x - \beta c t \\ y \\ z \end{pmatrix}$$

which is the Lorentz transformation given above. All tensors transform by the same rule.

The squared length of the differential of the position four-vector dx^{μ} constructed using

$$dx^2 = \eta_{\mu\nu} dx^{\mu} dx^{\nu} = -(c.dt)^2 + (dx)^2 + (dy)^2 + (dz)^2$$

is an invariant. Being invariant means that it takes the same value in all inertial frames, because it is a scalar (0 rank tensor), and so no Λ appears in its trivial transformation. Notice that when the line element dx^2 is negative that $d\tau = \sqrt{-dx^2}/c$ is the differential of proper time, while when dx^2 is positive, $\sqrt{dx^2}$ is differential of the proper distance.

The primary value of expressing the equations of physics in a tensor form is that they are then manifestly invariant under the Poincaré group, so that we do not have to do a special and tedious calculation to check that fact. Also in constructing such equations we often find that equations previously thought to be unrelated are, in fact, closely connected being part of the same tensor equation.

VELOCITY AND ACCELERATION IN 4D

Recognising other physical quantities as tensors also simplifies their transformation laws. First note that the velocity four-vector U^μ is given by

$$U^\mu = \frac{dx^\mu}{d\tau} = \begin{pmatrix} \gamma \\ \gamma v_x \\ \gamma v_y \\ \gamma v_z \end{pmatrix}$$

Recognising this, we can turn the awkward looking law about composition of velocities into a simple statement about transforming the velocity four-vector of one particle from one frame to another. U^μ also has an invariant form:

$$U^2 = \eta_{\nu\mu} U^\nu U^\mu = -c^2.$$

So all velocity four-vectors have a magnitude of c. This is an expression of the fact that there is no such thing as being at coordinate rest in relativity: at the least, you are always moving forward through time. The acceleration 4-vector is given by $A^\mu = dU^\mu / d\tau$. Given this, differentiating the above equation by τ produces

$$2\eta_{\mu\nu} A^\mu U^\nu = 0.$$

So in relativity, the acceleration four-vector and the velocity 4-vector are orthogonal.

MOMENTUM IN 4D

The momentum and energy combine into a covariant 4-vector:

$$p_\nu = m.\eta_{\nu\mu} U^\mu = \begin{pmatrix} -E \\ p_x \\ p_y \\ p_z \end{pmatrix}.$$

where m is the invariant mass.

The invariant magnitude of the momentum 4-vector is:

$$p^2 = \eta^{\mu\nu} p_\mu p_\nu = -(E/c)^2 + p^2.$$

We can work out what this invariant is by first arguing that, since it is a scalar, it doesn't matter which reference frame we calculate it, and then by transforming to a frame where the total momentum is zero.

We see that the rest energy is an independent invariant. A rest energy can be calculated even for particles and systems in motion, by translating to a frame in which momentum is zero.

The rest energy is related to the mass according to the celebrated equation discussed:

$$E_{rest} = mc^2$$

Note that the mass of systems measured in their centre of momentum frame (where total momentum is zero) is given by the total energy of the system in this frame. It may not be equal to the sum of individual system masses measured in other frames.

FORCE IN 4D

To use Newton's third law of motion, both forces must be defined as the rate of change of momentum with respect to the same time coordinate. That is, it requires the 3D force defined above. Unfortunately, there is no tensor in 4D which contains the components of the 3D force vector among its components.

If a particle is not traveling at c, one can transform the 3D force from the particle's co-moving reference frame into the observer's reference frame. This yields a 4-vector called the four-force. It is the rate of change of the above energy momentum four-vector with respect to proper time. The covariant version of the four-force is:

$$F_\nu = \frac{dp_\nu}{d\tau} = \begin{pmatrix} -dE/d\tau \\ dp_x/d\tau \\ dp_y/d\tau \\ dp_z/d\tau \end{pmatrix}$$

where τ is the proper time.

In the rest frame of the object, the time component of the four force is zero unless the "invariant mass" of the object is changing in which case it is the negative of that rate of change times c^2. In general, though, the components of the four force are not equal to the components of the three-force, because the three force is defined by the rate of change of momentum with respect to coordinate time, i.e. $\frac{dp}{dt}$ while the four force is defined by the rate of change of momentum with respect to proper time, i.e. $\frac{dp}{d\tau}$.

In a continuous medium, the 3D *density of force* combines with the *density of power* to form a covariant 4-vector. The spatial part is the result of dividing the force on a small cell (in 3-space) by the volume of that cell. The time component is the negative of the power transferred to that cell divided by the volume of the cell. This will be used below in the section on electromagnetism.

RELATIVITY AND UNIFYING ELECTROMAGNETISM

Theoretical investigation in classical electromagnetism led to the discovery of wave propagation. Equations generalizing the electromagnetic effects found that finite propagation-speed of the E and B fields required certain behaviours on charged particles. The general study of moving charges forms the Liénard–Wiechert potential, which is a step towards special relativity.

The Lorentz transformation of the electric field of a moving charge into a non-moving observer's reference frame results in the appearance of a mathematical term commonly called the magnetic field. Conversely, the *magnetic* field generated by a moving charge disappears and becomes a purely *electrostatic* field in a comoving frame of reference. Maxwell's equations are thus simply an empirical fit to special relativistic effects in a classical model of the Universe. As electric and magnetic fields are reference frame dependent and thus intertwined, one speaks of *electromagnetic* fields. Special relativity provides the transformation rules for how an electromagnetic field in one inertial frame appears in another inertial frame.

ELECTROMAGNETISM IN 4D

Maxwell's equations in the 3D form are already consistent with the physical content of special relativity. But we must rewrite them to make them manifestly invariant.

The charge density ρ and current density $[J_x, J_y, J_z]$ are unified into the current-charge 4-vector:

$$J^{\mu} = \begin{pmatrix} \rho \\ J_x \\ J_y \\ J_z \end{pmatrix}$$

The law of charge conservation becomes:

$$\partial_{\mu} J^{\mu} = 0.$$

The electric field $[E_x, E_y, E_z]$ and the magnetic induction $[B_x, B_y, B_z]$ are now unified into the (rank 2 antisymmetric covariant) electromagnetic field tensor:

$$F_{\mu\nu} = \begin{pmatrix} 0 & -E_x & -E_y & -E_z \\ E_x & 0 & B_z & -B_y \\ E_y & -B_z & 0 & B_x \\ E_z & B_y & -B_x & 0 \end{pmatrix}$$

The density of the Lorentz force f_μ exerted on matter by the electromagnetic field becomes:

$$f_\mu = F_{\mu\nu} J^\nu.$$

Faraday's law of induction and Gauss's law for magnetism combine to form:

$$\partial_\lambda F_{\mu\nu} + \partial_\mu F_{\nu\lambda} + \partial_\nu F_{\lambda\mu} = 0.$$

Although there appear to be 64 equations here, it actually reduces to just four independent equations. Using the antisymmetry of the electromagnetic field one can either reduce to an identity (0 = 0) or render redundant all the equations except for those with λ, μ, v = either 1,2,3 or 2,3,0 or 3,0,1 or 0,1,2.

The electric displacement [D_x, D_y, D_z] and the magnetic field [H_x, H_y, H_z] are now unified into the (rank 2 antisymmetric contravariant) electromagnetic displacement tensor:

$$D^{\mu\nu} = \begin{pmatrix} 0 & D_x & D_y & D_z \\ -D_x & 0 & H_z & -H_y \\ -D_y & -H_z & 0 & H_x \\ -D_z & H_y & -H_x & 0 \end{pmatrix}$$

Ampere's law and Gauss's law combine to form:

$$\partial_\nu D^{\mu\nu} = j^\mu.$$

In a vacuum, the constitutive equations are:

$$\mu_0 D^{\mu\nu} = \eta^{\mu\alpha} \eta^{\nu\beta} F_{\alpha\beta}.$$

Antisymmetry reduces these 16 equations to just six independent equations.

The energy density of the electromagnetic field combines with Poynting vector and the Maxwell stress tensor to form the 4D electromagnetic stress-energy tensor. It is the flux (density) of the momentum 4-vector and as a rank 2 mixed tensor it is:

$$T_{\alpha}^{\pi} = F_{\alpha\beta}D^{\pi\beta} - \frac{1}{4}\delta_{\alpha}^{\pi}F_{\mu\nu}D^{\mu\nu}$$

where δ_{α}^{π} is the Kronecker delta. When upper index is lowered with η, it becomes symmetric and is part of the source of the gravitational field.

The conservation of linear momentum and energy by the electromagnetic field is expressed by:

$$f_{\mu} + \partial_{\nu}T_{\mu}^{\nu} = 0$$

where f_{μ} is again the density of the Lorentz force. This equation can be deduced from the equations above (with considerable effort).

TENSOR

The term tensor has slightly different meanings in mathematics and physics. In the mathematical fields of multilinear algebra and differential geometry, a tensor is a multilinear function. In physics and engineering, the same term usually means what a mathematician would call a tensor field: an association of a different (mathematical) tensor with each point of a geometric space, varying continuously with position.

History

The word *tensor* was introduced in 1846 by William Rowan Hamilton to describe the norm operation in a certain type of algebraic system (eventually known as a Clifford algebra). The word was used in its current meaning by Woldemar Voigt in 1899.

Tensor calculus was developed around 1890 by Gregorio Ricci-Curbastro under the title *absolute differential calculus*, and was made accessible to many mathematicians by the publication of Tullio Levi-Civita's 1900 classic text of the same name (in Italian; translations followed). In the 20th century, the subject came to be known as *tensor analysis*, and achieved broader acceptance with the introduction of Einstein's theory of general relativity, around 1915.

General relativity is formulated completely in the language of tensors. Einstein had learned about them, with great difficulty, from the geometer Marcel Grossmann, or perhaps from Levi-Civita himself. Tensors are used also in other fields such as continuum mechanics.

TWO USAGES OF 'TENSOR'

In mathematics, a tensor is (in an informal sense) a generalized linear 'quantity' or 'geometrical entity' that can be expressed as a multi-dimensional array relative to a choice of basis of the particular space on which it is defined. The intuition underlying the tensor concept is inherently geometrical: as an object in and of itself, a tensor is *independent of any chosen frame of reference*. However, in the modern treatment, tensor theory is best regarded as a topic in multilinear algebra. Engineering applications do not usually require the full, general theory, but theoretical physics now does.

For example, the Euclidean inner product (dot product) a real-valued function of two vectors that is linear in each is a mathematical tensor. Similarly, on a smooth curved surface such as a torus, the metric tensor (field) essentially defines a different inner product of tangent vectors at each point of the surface. Just as a linear transformation can be represented as a matrix of numbers with respect to given vector bases, so a tensor can be written as an organized collection of numbers. In physics, the numbers may be obtained as physical quantities that depend on a basis, and the collection is determined to be a tensor if the quantities transform appropriately under change of basis.

Physical - tensor Fields

Many mathematical structures informally called 'tensors' are actually 'tensor fields' – an abstraction of tensors to field, wherein tensorial quantities vary from point to point. Differential equations posed in terms of tensor quantities are basic to modern mathematical physics, so that methods of differential calculus are also applied to tensors.

Tensor rank

The rank of a particular tensor is the number of array indices required to describe such a quantity. For example in classical mechanics, mass, temperature, and other scalar quantities are tensors of rank 0; but force, momentum and other vector-like quantities are tensors of rank 1. The novel aspects of tensor theory are seen from rank 2 onwards: A linear transformation such as an anisotropic relationship (relativistic mass) between force and acceleration vectors is a tensor of rank. There exists a similar relationship between temperature in different reference frames in relativistic thermodynamics.

Tensor Valence

In physical applications, array indices are distinguished by being contravariant (superscripts) or covariant (subscripts), depending upon the type of transformation properties. The valence of a particular tensor is the number and type of array indices; tensors with the same rank but different valence are not, in general, identical. However, any given covariant index can be transformed into a contravariant one, and vice versa, by applying the metric tensor. This operation is generally known as raising or lowering indices.

Importance and Applications

Tensors are important in physics and engineering. In the field of diffusion tensor imaging, for instance, a tensor quantity that expresses the differential permeability of organs to water in varying directions is used to produce scans of the brain; in this technique tensors are in effect made visible. Perhaps the most important engineering examples are the stress tensor and strain tensor, which are both 2nd rank tensors, and are related in a general linear material by a fourth rank elasticity tensor.

Specifically, a 2nd rank tensor quantifying stress in a 3-dimensional/solid object has components which can be conveniently represented as a 3x3 array. The three Cartesian faces of a cube-shaped infinitesimal volume segment of the solid are each subject to some given force. The force's vector components are also three in number (being in three-space). Thus, 3x3, or 9 components are required to describe the stress at this cube-shaped infinitesimal segment (which may now be treated as a point). Within the bounds of this solid is a whole mass of varying stress quantities, each requiring 9 quantities to describe. Thus, the need for a 2nd order tensor is produced.

While tensors can be represented by multi-dimensional arrays of components, the point of having a tensor *theory* is to explain further implications of saying that a quantity is a *tensor*, beyond specifying that it requires a number of indexed components. In particular, tensors behave in specific ways under coordinate transformations. The abstract theory of tensors is a branch of linear algebra, now called multilinear algebra.

THE CHOICE OF APPROACH

There are two ways of approaching the definition of tensors:

- The usual physics way of defining tensors, in terms of objects whose components transform according to certain rules, introducing the ideas of covariant or contravariant transformations.
- The usual mathematics way, which involves defining certain vector spaces and not fixing any coordinate systems until bases are introduced when needed. Covariant vectors, for instance, can also be described

as one-forms, or as the elements of the dual space to the contravariant vectors.

Physicists and engineers are among the first to recognise that vectors and tensors have a physical significance as entities, which goes beyond the (often arbitrary) co-ordinate system in which their components are enumerated. Similarly, mathematicians find there are some tensor relations which are more conveniently derived in a co-ordinate notation.

Examples

Physical Examples

As a simple example, consider a ship in the water. We want to describe its response to an applied force. Force is a vector, and the ship will respond with an acceleration, which is also a vector. The relationship between force and acceleration is linear in classical mechanics. Such a relationship is described by a rank two tensor of type that is to say, here it transforms a plane vector into another such vector. The tensor can be represented as a matrix which when multiplied by a vector results in another vector. Just as the numbers which represent a vector will change if one changes the coordinate system, the numbers in the matrix that represents the tensor will also change when the coordinate system is changed.

In engineering, the stresses inside a solid body or fluid are also described by a tensor; the word "tensor" is Latin for something that stretches, i.e., causes tension. If a particular surface element inside the material is singled out, the material on one side of the surface will apply a force on the other side. In general, this force will not be orthogonal to the surface, but it will depend on the orientation of the surface in a linear manner. This is described by a tensor of type (2,0), in linear elasticity, or more precisely by a tensor *field* of type (2,0) since the stresses may change from point to point.

Mathematical Examples

Some well-known examples of tensors in differential geometry are quadratic forms, such as metric tensors, and the curvature tensor.

Formally speaking, a tensor has a particular type according to the construction with tensor products that give rise to it. For computational purposes, it may be expressed as the sequence of values represented by a function with a tuple-valued domain and a scalar valued range. Domain values are tuples of counting numbers, and these numbers are called indices. For example, a rank 3 tensor might have dimensions 2, 5, and 7. Here, the indices range from «1, 1, 1» through «2, 5, 7»; thus the tensor would have one value at «1, 1, 1», another at «1, 1, 2», and so on for a total of 70 values. As a special case, (finite-dimensional) vectors may be expressed as a sequence of values represented by a function with a scalar valued domain and a scalar valued range; the number of distinct indices is the dimension of the vector. Using this approach, the rank 3 tensor of dimension (2,5,7) can be represented as a 3-dimensional array of size 2 × 5 × 7. In this usage, the number of "dimensions" comprising the array is equivalent to the "rank" of the tensor, and the dimensions of the tensor are equivalent to the "size" of each array dimension.

A tensor field associates a tensor value with every point on a manifold. Thus, instead of simply having 70 values as indicated in the above example, for a rank 3 tensor field with dimensions «2, 5, 7»; every point in the space would have 70 values associated with it. In other words, a tensor field means there's some tensor-valued function which has, for example, Euclidean space as its domain.

Approaches, in Detail

There are ***equivalent*** approaches to visualizing and working with tensors; that the content is actually the same may only become apparent with some familiarity with the material.

- *The classical approach:* The classical approach views tensors as multidimensional arrays that are n-dimensional generalizations of scalars, 1-dimensional vectors and 2-dimensional matrices. The "components" of the tensor are the values in the array. This idea can then be further generalized to tensor fields, where the elements of the tensor are functions, or even differentials.

However, to count as a tensor, the arrays need to transform correctly when the reference co-ordinate system is changed. This transformation is a generalisation of the relationship which holds for vector components, and is similarly an expression of the independence of the underlying entity from the reference frame in which it is expressed.

- *The modern approach:* The modern (component-free) approach views tensors initially as abstract objects, expressing some definite type of multi-linear concept. Their well-known properties can be derived from their definitions, as linear maps or more generally; and the rules for manipulations of tensors arise as an extension of linear algebra to multilinear algebra. This treatment has attempted to replace the component-based treatment for advanced study, in the way that the more modern component-free treatment of vectors replaces the traditional component-based treatment after the component-based treatment has been used to provide an elementary motivation for the concept of a vector. You could say that the slogan is 'tensors are elements of some tensor space'. Nevertheless, a component-free approach has not become fully popular, owing to the difficulties involved with giving a geometrical interpretation to higher-rank tensors.
- The intermediate treatment of tensors article attempts to bridge the two extremes, and to show their relationships.

In the end the same computational content is expressed, both ways.

TENSOR DENSITIES

It is also possible for a tensor field to have a "density". A tensor with density r transforms as an ordinary tensor under coordinate transformations, except that it is also multiplied by the determinant of the Jacobian to the rth power. Invariantly, in the language of multilinear algebra, one can think of tensor densities as multilinear maps taking their values in the (1-dimensional) space of n-forms (where n is the dimension of the space), as opposed to taking their values in just R. Higher "weights" then just correspond to taking additional tensor products with this space in the range. In the language of vector bundles, the determinant bundle of the tangent bundle is a line bundle that can be used to 'twist' other bundles r times.

CONTRAVARIANT AND COVARIANT INDICES

Rank and Dimension

Now that we have talked about tensors, we need to figure out how to classify them. One important characteristic is the rank of a tensor, which is the number of indices needed to specify the tensor. An ordinary matrix is a rank 2 tensor, a vector is a rank 1 tensor, and a scalar is rank 0. Tensors can, in general, have rank greater than 2, and often do.

Another characteristic of a tensor is the dimension of the tensor, which is the count of each index. For example, if we have a matrix consisting of 3 rows, with 4 elements in each row (columns), then the matrix is a tensor of dimension (3,4), or equivalently, dimension.

The important thing about rank and dimension is that they are invariant to changes in the coordinate system. You can change the coordinate system all you want, and the rank and the dimensions don't change. This brings up the important question of how tensors do change when you change the coordinate system. One thing we shall find when we look at the question is that in reality there are two different types of vectors.

Contravariant and Covariant Vectors

Imagine that you are driving a car at 100 kilometers per hour toward the northwest. Lets call this vector v. Suddenly you realise that you are in a meter-ish mood and so we want to figure out how fast you are going using meters instead of kilometers. Quickly changing your coordinate system, you find that you are travelling 100 * 1000 meters per hour toward the northwest. No problem.

Now you are in the rolling countryside, and you notice the temperature changing. We then draw a map of how the temperature changes as we move across the countryside. We then travel along the path of steepest descent. We notice at a given point that the temperature is changing at 10 Celsius degrees per kilometer toward the southwest. Let's call this vector w. Again you go into a meter-ish mood. Doing a quick calculation you figure out that the gradient of the temperature change is downward at 10/1000 Celsius degrees per meter.

Even though we are talking about two vectors we are treating them very differently when we change our coordinates. In the first case, the vector reacted to the coordinate change by a multiplication. In the second case, we did a division. The first case we were changing a vector that was distance per something, while in the second case, the vector was something per distance. These are two very different types of vectors.

The mathematical term for the first type of vector is called a contravariant vector. The second type of vector is called a covariant vector. Sometimes a covariant vector is called a one form.

Attempting a Fuller Explanation

It is easy to see why w is called covariant. Covariant simply means that the characteristic that w measures, change in temperature, increases in magnitude with an increase in displacement along the coordinate system. In other words, the further you travel from a fixed point, the more the temperature changes, or equivalently, change in temperature *covaries* with change in displacement.

Although it is a bit more difficult to see, v is called *contravariant* for precisely the *opposite* reason. Since v represents a velocity, or *distance per unit time,* we can think of v as the inverse of *time per unit distance,* meaning the amount of time that passes in traveling a certain fixed amount of distance.

Time per unit distance is clearly *covariant*, because as you travel further and further from a fixed point, more and more time elapses. In other words, time covaries with displacement. Since velocity is the *inverse* of time per unit distance, than it follows that velocity must be *contravariant*.

The difference is also evident in the units of measure. The units of measure for v are *meters per hour*, whereas the units for w are *degrees Celsius per meter*. The coordinate system is position in space, measured in units of *meters*. So again, we see that the coordinate system appears in the numerator of v, which suggests that v is *contravariant* (with inverse time in this case), wherease the coordinate system appears in the denominator of w, which indicates that w is *covariant* (with change in temperature).

Contravariant vectors describe those quantities where the distance units comes at the numerator (like *velocity*), whereas covariant are those where the distance unit is at the denominator (like *temperature gradient*).

These are, of course, just fancy mathematical names. As we can see contravariant vectors and covariant vectors are very different from each other and we want to avoid confusing them with each other. To do this mathematicians have come up with a clever notation. The components of a contravariant vector are represented by superscripts, while the components of a covariant vector are represented by subscripts. So the components of vector v are v^1 and v^2 while the components of vector w are w_1 and w_2.

SCALE INVARIANCE

Now that we have contravariant vectors and covariant vectors, we can do something very interesting and combine them. We have a contravariant vector that describes the direction and speed at which we are going. We have covariant vector that describes the rate and direction at which the temperature changes. If we combine them using the dot product

$$f = v.\,w$$

we get the rate at which the temperature changes, *f*, as we move in a certain direction, with units of degrees Celsius per second. The interesting thing about the units of *f* is that they do *not* include any units of distance, such as meters or kilometers. So now suppose we *change* the coordinate system from meters to kilometers. How does *f* change? It doesn't. We call this characteristic scale invariance, and we say that *f* is a scale invariant quantity. The value of *f* is *invariant* with changes in the *scale* of the coordinate system.

Now so far we have been treating w as if it were just an odd type of vector. But there is a another more powerful way of thinking about *w*. Look at what we just did. We took *v*, combined it with w and got something that doesn't change when you change the coordinate system. Now one way of thinking about it is to say that *w* is a function, that takes *v* and converts it into a scale invariant value, *f*.

Vector Spaces and Basis Vectors

This fact, that a covariant vector like *w* can convert any contravariant vector like v into a scale invariant value like *f*, is described by saying that w is a linear functional.

Can we be more precise about using the word *like*? Yes, and it's important. Mathematical operations such as *converting one sort of vector into another sort of vector* are done on vector spaces. See vector space for a careful definition of vector spaces. Here, loosely speaking, let us say that a vector space is a set of vectors which can be added together and multiplied by numbers and that the result is always another vector in the same vector space.

Let us define *V* to be the vector space of contravariant vectors like *v*.

Then, the set of all covariant vectors like w, which convert vectors like v from *V* into scalars like *f*, which we can also call the set of all linear functionals w on *V*, can be given the name V^*, which we call the dual space.

V^* is also a vector space.

Now we can be more careful about the word *like*, by saying which spaces each object must be a member of: any vector w in V^* (called a **covariant** vector, or a **1-form**) can convert any vector v in V (called a **contravariant** vector) into a scale invariant value like f. (We have not said what set f is a member of: in practice, we will usually only be interested in f as a member of the set of real numbers.)

Any vector space has a set of basis vectors. Components of contravariant vectors are written with superscript ("upper") indices, but the basis vectors are written with subscript ("lower") indices. If the set $\{e_\alpha\}$ is a basis for V, then $v \in V$ is written as the linear combination $v = v^\mu e_\mu$. (We are using Einstein summation notation, detailed in the next section; this is shorthand for $\sum_\mu v^\mu e_\mu$.)

Before moving on to covariant vectors, we must define the notion of a dual basis. Remember that elements of V^* are linear functionals on V. So we can "apply" covariant vectors to contravariant vectors to get a scalar. For example, if $\sigma \in V^*$ and $v \in V$, then $\sigma(v)$ returns a scalar. Now, the dual basis is defined as follows: if $\{e_\alpha\}$ is a basis for V, then the dual basis is a basis $\{\omega^\alpha\}$ for V^* which satisfies $\omega^\mu(e_\nu) = \delta^\mu_\nu$ (where δ^μ_ν is the Kronecker delta) for every μ and v. Note that the dual basis for the canonical basis is usually written as $\{dx^\alpha\}$, for reasons we will not go into in this section.

Now, the components of covariant vectors are written with subscript ("lower") indices. As $\{\omega^\alpha\}$ is a basis for V^*, we can write a covariant vector σ as $\sigma = \sigma_\mu \omega^\mu$.

We can now evaluate any functional (covariant vector) applied to any vector (contravariant vector). If $\sigma = V^*$ and $v \in$

V, then by linearity $\sigma(v) = \sigma_\alpha \omega^\alpha (v^\beta e_\beta) = \sigma_\alpha v^\beta \omega^\alpha (e_\beta)$ $= \sigma_\alpha v^\beta \delta^\alpha_\beta = \sigma_\alpha v^\alpha$. Finally, if we define $e_\alpha(\omega^\beta) = \delta^\beta_\alpha$, we see that $v(\omega) = \omega(v)$.

CONTRAVARIANT

One of the most important relations involving continuous functions of multiple continuous variables (such as coordinates) is the formula for the *total differential*. In general if we are given a smooth continuous function $y = f(x^1, x^2, \ldots, x^n)$ of n variables, the incremental change dy in the variable y resulting from incremental changes $dx^1, dx^2, \ldots, dx^n$ in the variables $x^1, x^2, \ldots, x^n$ is given by

$$dy = \frac{\partial y}{\partial x^1} dx^1 + \frac{\partial y}{\partial x^2} dx^2 + \ldots + \frac{\partial y}{\partial x^n} dx^n$$

where $\partial y / \partial x^i$ is the partial derivative of y with respect to x^i. (The superscripts on x are just indices, not exponents.) The scalar quantity dy is called the total differential of y. This formula just expresses the fact that the total incremental change in y equals the sum of the "sensitivities" of y to the independent variables multiplied by the respective incremental changes in those variables. (See the Appendix for a slightly more rigorous definition.)

If we define the vectors

$$g = \left[\frac{\partial y}{\partial x^1}, \frac{\partial y}{\partial x^2}, \ldots, \frac{\partial y}{\partial x^n}\right] \quad \mathrm{d} = [dx^1, dx^2, \ldots, dx^n]$$

then dy equals the scalar (dot) product of these two vectors, i.e., we have $dy = g \times d$. Regarding the variables $x^1, x^2, \ldots, x^n$ as coordinates on a manifold, the function $y = f(x^1, x^2, \ldots, x^n)$ defines a scalar field on that manifold, g is the gradient of y (often denoted as ∇y), and d is the differential position of x (often denoted as dx), all evaluated about some nominal point $[x^1, x^2, \ldots, x^n]$ on the manifold.

The gradient $g = \nabla y$ is an example of a *covariant tensor,* and the differential position $d = dx$ is an example of a *contravariant tensor.* The difference between these two kinds of tensors is how they transform under a continuous change of coordinates. Suppose we have another system of smooth continuous coordinates $X^1, X^2, \ldots, X^n$ defined on the same manifold. Each of these new coordinates can be expressed (in the region around any particular point) as a function of the original coordinates, $X^i = F^i(x^1, x^2, \ldots, x^n)$, so the total differentials of the new coordinates can be written as

$$dX^i = \frac{\partial X^i}{\partial x^1} dx^1 + \frac{\partial X^i}{\partial x^2} dx^2 + \ldots + \frac{\partial X^i}{\partial x^n} dx^n$$

Thus, letting D denote the vector $[dX^1, dX^2, \ldots, dX^n]$ we see that the components of D are related to the components of d by the equation

$$D^i = \frac{\partial X^i}{\partial x^1} d^1 + \frac{\partial X^i}{\partial x^2} d^2 + \ldots + \frac{\partial X^i}{\partial x^n} d^n$$

This is the prototypical transformation rule for a contravariant tensor of the first order. On the other hand, the gradient vector $g = \nabla y$ is a covariant tensor, so it doesn't transform in accord with this rule. To find the correct transformation rule for the gradient (and for covariant tensors in general), note that if the system of functions F^i is invertible (which it is if and only if the determinant of the Jacobian is non-zero), then the original coordinates can be expressed as some functions of these new coordinates, $x^i = f^i(X^1, X^2, \ldots, X^n)$ for $i = 1, 2, \ldots, n$. This enables us to write the total differentials of the original coordinates as

$$dx^i = \frac{\partial x^i}{\partial X^1} dX^1 + \frac{\partial x^i}{\partial X^2} dX^2 + \ldots + \frac{\partial x^i}{\partial X^n} dX^n$$

If we now substitute these expressions for the total coordinate differentials into equation and collect by differentials of the new coordinates, we get

$$dy=\sum_{k=1}^{n}\left(\frac{\partial x^1}{\partial X^k}\frac{\partial y}{\partial x^1}+\frac{\partial x^2}{\partial X^k}+...+\frac{\partial x^n}{\partial X^k}\frac{\partial y}{\partial x^n}\right)dX^k$$

Thus, the components of the gradient of g of y with respect to the X^i coordinates are given by the quantities in parentheses. If we let G denote the gradient of y with respect to these new coordinates, we have

$$G_i=\frac{\partial x^1}{\partial X^i}g_1+\frac{\partial x^2}{\partial X^i}g_2+...+\frac{\partial x^n}{\partial X^i}g_n$$

This is the prototypical transformation rule for covariant tensors of the first order. Comparing this with the contravariant rule givene, we see that they both define the transformed components as linear combinations of the original components, but in the contravariant case the coefficients are the partials of the new coordinates with respect to the old, whereas in the covariant case the coefficients are the partials of the old coefficients with respect to the new.

The key attribute of a tensor is that it's representations in different coordinate systems depend only on the relative orientations and scales of the coordinate axes at that point, not on the absolute values of the coordinates. This is why the absolute position vector pointing from the origin to a particular object in space is not a tensor, because the components of its representation depend on the absolute values of the coordinates. In contrast, the coordinate differentials transform based solely on local information.

So far we have discussed only first-order tensors, but we can define tensors of any order. One of the most important examples of a second-order tensor is the metric tensor. Recall that the generalized Pythagorean theorem enables us to express the squared differential distance ds along a path on the spacetime manifold to the corresponding differential components dt, dx, dy, dz as a general quadratic function of those differentials as follows

$$(ds)^2 = g_{00}(dt)(dt) + g_{01}(dt)(dx) + g_{02}(dt)(dy) + g_{03}(dt)(dz)$$

$$+ g_{10}(dx)(dt) + g_{11}(dx)(dx) + g_{12}(dx)(dy) + g_{13}(dx)(dz)$$
$$+ g_{20}(dy)(dt) + g_{21}(dy)(dx) + g_{22}(dy)(dy) + g_{23}(dz)(dz)$$
$$+ g_{30}(dz)(dt) + g_{31}(dz)(dx) + g_{32}(dz)(dy) + g_{33}(dz)(dz)$$

Naturally if we set $g_{00} = -g_{11} = -g_{22} = -g_{33} = 1$ and all the other g_{ij} coefficients to zero, this reduces to the Minkowski metric. However, a different choice of coordinate systems requires the use of the full formula. To simplify the notation, it's customary to use the indexed variables x^0, x^1, x^2, x^3 in place of t, x, y, z respectively. This allows us to express the above metrical relation in abbreviated form as

$$(dx)^2 = \sum_{\mu=0}^{3}\sum_{\nu=0}^{3} g_{\mu\nu}dx^{\mu}dx^{\nu}$$

To abbreviate the notation even more, we adopt Einstein's convention of omitting the summation symbols altogether, and simply stipulating that summation from 0 to 3 is implied over any index that appears more than once in a given product. With this convention the above expression is written as

$$(ds)^2 = g_{\mu\nu}dx^{\mu}dx^{\nu}$$

Notice that this formula expresses something about the intrinsic metrical relations of the space, but it does so in terms of a specific coordinate system. If we considered the metrical relations at the same point in terms of a different system of coordinates (such as changing from Cartesian to polar coordinates), the coefficients $g_{\mu\nu}$ would be different. Fortunately there is a simple way of converting the $g_{\mu\nu}$ from one system of coordinates to another, based on the fact that they describe a purely localistic relation among differential quantities. Suppose we are given the metrical coefficients $g_{\mu\nu}$ for the coordinates x^{α}, and we are also given another system of coordinates y^{α} that are defined in terms of the x^{α} by some arbitrary continuous functions

$y^0 = f_0(x^0, x^1, x^2, x^3)$ $y^1 = f_1(x^0, x^1, x^2, x^3)$

$y^2 = f_2(x^0, x^1, x^2, x^3)$ $y^3 = f_3(x^0, x^1, x^2, x^3)$

Assuming the Jacobian of this transformation isn't zero, we know that it's invertible, and so we can just as well express the original coordinates as continuous functions (at this point) of the new coordinates

$x^0 = F_0(y^0, y^1, y^2, y^3)$ $x^1 = F_1(y^0, y^1, y^2, y^3)$

$x^0 = F_2(y^0, y^1, y^2, y^3)$ $x^3 = F_3(y^0, y^1, y^2, y^3)$

Now we can evaluate the total derivatives of the original coordinates in terms of the new coordinates. For example, dx^0 can be written as

$$dx^0 = \frac{\partial x^0}{\partial y^0} dy^0 + \frac{\partial x^0}{\partial y^1} dy^1 + \frac{\partial x^0}{\partial y^0} dy^2 + \frac{\partial x^0}{\partial y^3} dy^3$$

and similarly for the dx^1, dx^2, and dx^3. The product of any two of these differentials, dx^μ and dx^v, is of the form

$$dx^\mu dx^\nu = \frac{\partial x^\mu}{\partial y^\alpha} \frac{\partial x^\nu}{\partial y^\beta} dy^\alpha dy^\beta$$

(remembering the summation convention). Substituting these expressions for the products of x differentials in the metric formula gives

$$(ds)^2 = g_{\mu\nu} \frac{\partial x^\mu}{\partial y^\alpha} \frac{\partial x^\nu}{\partial y^\beta} dy^\alpha dy^\beta$$

The first three factors on the right hand side obviously represent the coefficient of $dy^\alpha dy^\beta$ in the metric formula with respect to the y coordinates, so we've shown that the array of metric coefficients transforms from the x to the y coordinate system according to the equation

$$(g_{\alpha\beta})_y = \frac{\partial x^\mu}{\partial y^\alpha} \frac{\partial x^\nu}{\partial y^\beta} (g_{\mu\nu})_x$$

Notice that each component of the new metric array is a linear combination of the old metric components, and the coefficients are the partials of the old coordinates with respect to the new. Arrays that transform in this way are called covariant tensors.

On the other hand, if we define an array $A^{\mu\nu}$ with the components $(dx^{\mu}/ds)(dx^{\nu}/ds)$ where s denotes a path parameter along some particular curve in space, then equation tells us that this array transforms according to the rule.

$$(A^{\mu\nu})_x = \frac{\partial x^{\mu}}{\partial y^{\alpha}} \frac{\partial x^{\nu}}{\partial y^{\beta}} (A^{\alpha\beta})_y$$

This is very similar to the previous formula, except that the partial derivatives are of the new coordinates with respect to the old. Arrays whose components transform according to this rule are called contra-variant tensors.

When we speak of an array being transformed from one system of coordinates to another, it's clear that the array must have a definite meaning independent of the system of coordinates. We could, for example, have an array of scalar quantities, whose values are the same at a given point, regardless of the coordinate system. However, the components of the array might still be required to change for different systems. For example, suppose the temperature at the point (x, y, z) in a rectangular tank of water is given by the scalar field $T(x, y, z)$, where x, y, z are Cartesian coordinates with origin at the geometric centre of the tank. If we change our system of coordinates by moving the origin, say, to one of the corners of the tank, the function $T(x, y, z)$ must change to $T(x - x_0, y - y_0, z - z_0)$. But at a given physical point the value of T is unchanged.

Notice that g_{20} is the coefficient of $(dy)(dt)$, and g_{02} is the coefficient of $(dt)(dy)$, so without loss of generality we could combine them into the single term $(g_{20} + g_{02})(dt)(dy)$. Thus the individual values of g_{20} and g_{02} are arbitrary for a given metrical equation, since all that matters is the sum $(g_{20} + g_{02})$. For this reason we're free specify each of those coefficients as half the sum, which results in $g_{20} = g_{02}$. The same obviously applies to all the other diagonally symmetric pairs, so for the sake of definiteness and simplicity we can set $g_{ab} = g_{ba}$. It's important to note, however, that this symmetry property doesn't apply to all tensors. In general we have no *a priori* knowledge of the symmetries of an arbitrary tensor.

Incidentally, when we refer to a vector (or, more generally, a tensor) as being either contravariant or covariant we're abusing the language slightly, because those terms really just signify two different conventions for interpreting the *components* of the object with respect to a given coordinate system, whereas the essential attributes of a vector or tensor are independent of the particular coordinate system in which we choose to express it. In general, any given vector or tensor can be expressed in both contravariant and covariant form with respect to any given coordinate system. For example, consider the vector *P* is shown.

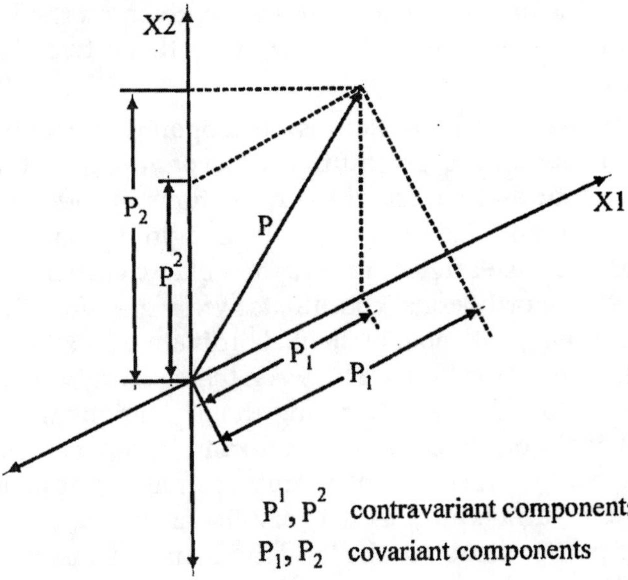

P^1, P^2 contravariant components
P_1, P_2 covariant components

We should note that when dealing with a vector field on a manifold each element of the field exists entirely at a single point of the manifold, with a direction and a magnitude, rather than imagining each vector to actually extends from one point in the manifold to another. However, for the purpose of illustrating the relation between contravariant and covariant components, we are focusing on simple displacement vectors in a flat metrical space, which can be considered to extend from one point to another.

Figure shows an arbitrary coordinate system with the axes X1 and X2, and the contravariant and covariant components of the position vector P with respect to these coordinates. As can be seen, the jth contravariant component consists of the projection of P onto the jth axis parallel to the other axis, whereas the jth covariant component consists of the projection of P into the jth axis perpendicular to that axis. This is the essential distinction (up to scale factors) between the contravariant and covariant ways of expressing a vector or, more generally, a tensor. It may seem that the naming convention is backwards, because the "contra" components go *with* the axes, whereas the "co" components go *against* the axes, but historically these names were given on the basis on the transformation laws that apply to these two different interpretations.

If the coordinate system is "orthogonal" (meaning that the coordinate axes are mutually perpendicular) then the contravariant and covariant interpretations are identical (up to scale factors). This can be seen by imagining that we make the coordinate axes in Figure perpendicular to each other. Thus when we use orthogonal coordinates we are essentially using both contravariant and covariant coordinates, because in such a context the only difference between them (at any given point) is scale factors. It's worth noting that "orthogonal" doesn't necessarily imply "rectilinear". For example, polar coordinates are not rectilinear, i.e., the axes are not straight lines, but they *are* orthogonal, because as we vary the angle we are always moving perpendicular to the local radial axis. Thus the metric of a polar coordinate system is diagonal, just as is the metric of a Cartesian coordinate system, and so the contravariant and covariant forms at any given point differ only by scale factors (although these scale factor may vary as a function of position). Only when we consider systems of coordinates that are not mutually perpendicular do the contravariant and covariant forms differ (at a given point) by more than just scale factors.

To understand in detail how the representations of vectors in different coordinate systems are related to each other, consider the displacement vector P in a flat 2-dimensional space shown.

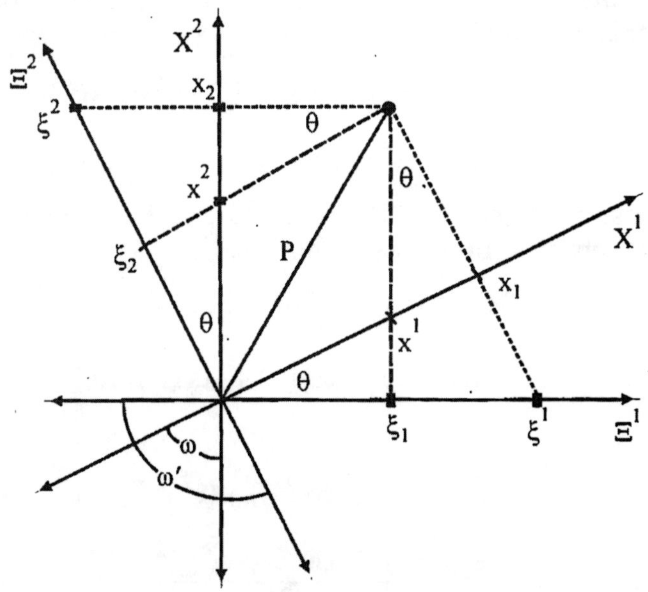

In terms of the X coordinate system the contravariant components of P are (x^1, x^2) and the covariant components are (x_1, x_2). We've also shown another set of coordinate axes, denoted by Ξ, defined such that Ξ^1 is perpendicular to X^2, and X^2 is perpendicular to X^1. In terms of these alternate coordinates the contravariant components of P are (ξ^1, ξ^2) and the covariant components are (ξ_1, ξ_2). The symbol w signifies the angle between the two positive axes X^1, X^2, and the symbol ω' denotes the angle between the axes Ξ^1 and Ξ^2. These angles satisfy the relations $\omega + \omega' = \pi$ and $\theta = (\omega' - \omega)/2$. We also have

$$x_1 = \xi^1 \cos(\theta) \quad x_2 = \xi^2 \cos(\theta)$$

$$\xi_1 = x^1 \cos(\theta) \quad \xi_2 = x^2 \cos(\theta)$$

which shows that the covariant components with respect to the X coordinates are the same, up to a scale factor of cos(θ), as the contravariant components with respect to the Ξ coordinates, and vice versa. For this reason the two coordinate systems are called "duals" of each other. Making use of the additional relations

$$x^1 - x^1 = x^2 \sin(\theta) \quad \xi^1 - \xi_1 = \xi^2 \sin(\theta)$$

$$x^2 - x^2 = x^1 \sin(\theta) \quad \xi^2 - \xi_2 = \xi^1 \sin(\theta)$$

the squared length of P can be expressed in terms of any of these sets of components as follows:

$$s^2 = (x^1)^2 + 2x^1x^2 \cos(\omega) + (x^2)^2$$

$$= \frac{(x_1)^2 - 2x_1x_2 \cos(\omega) + (x_2)^2}{\sin(\omega)^2}$$

$$= (\xi^1)^2 + 2\xi^1\xi^2 \cos(\omega') + (\xi^2)^2$$

$$= \frac{(\xi_1)^2 - 2\xi_1\xi_2 \cos(\omega') + (\xi_2)^2}{\sin(\omega')^2}$$

In general the squared length of an arbitrary vector on a (flat) 2-dimensional surface can be given in terms of the contravariant components by an expression of the form

$$s^2 = g_{11}x^1x^1 + g_{12}x^1x^2 + g_{21}x^2x^1 + g_{22}x^2x^2$$

where the coefficients g_{uv} are the components of the covariant *metric tensor*. This tensor is always symmetrical, meaning that $g_{uv} = g_{uv}$, so there are really only three independent elements for a two-dimensional manifold. With Einstein's summation convention we can express the preceding equation more succinctly as

$$s^2 = g_{uv}x^ux^v$$

From the preceding formulas we can see that the covariant metric tensor for the X coordinate system in Figure is

$$(g_{uv})_X = \begin{bmatrix} 1 & \cos(\omega) \\ \cos(\omega) & 1 \end{bmatrix}$$

whereas for the dual coordinate system Ξ the covariant metric tensor is

$$(g_{uv})_\Xi \begin{bmatrix} 1 & -\cos(w) \\ -\cos(w) & 1 \end{bmatrix}$$

noting that $\cos(\omega') = -\cos(\omega)$. The determinant g of each of these matrices is $\sin(\omega)^2$, so we can express the relationship between the dual systems of coordinates as

$$x_u = \xi^u \sqrt{g} \quad x^u = \xi_u \frac{1}{\sqrt{g}}$$

We will find that the *inverses* of the metric tensor is also very useful, so let's use the superscripted symbol g^{uv} to denote the inverse of a given g_{uv}. The inverse metric tensors for the X and Ξ coordinate systems are

$$(g^{uv})_X = \frac{1}{\sin(\omega)^2} \begin{bmatrix} 1 & -\cos(\omega) \\ -\cos(\omega) & 1 \end{bmatrix}$$

$$(g^{uv})_\Xi = \frac{1}{\sin(\omega)^2} \begin{bmatrix} 1 & \cos(\omega) \\ \cos(\omega) & 1 \end{bmatrix}$$

Comparing the left-hand matrix with the previous expression for s^2 in terms of the covariant components, we see that

$$x^2 = g^{uv} x_u x_v$$

so the inverse of the covariant metric tensor is indeed the contravariant metric tensor. Now let's consider a vector x whose contravariant components relative to the X axes of Figure are x^1, x^2, and let's multiply this by the covariant metric tensor as follows:

$$x_v = g_{uv} x^u$$

Remember that summation is implied over the repeated index u, whereas the index v appears only once (in any given product) so this expression applies for any value of v. Thus the expression represents the two equations

$$x_1 = g_{11}x^1 + g_{21}x^2$$
$$x_2 = g_{12}x^1 + g_{22}x^2$$

If we carry out this multiplication we find

$$x_1 = x^1 + x^2 \cos(\omega)$$

$$x_2 = x^1 \cos(\omega) + x^2$$

which agrees with the previously stated relations between the covariant and contravariant components, noting that $\sin(\theta) = \cos(\omega)$. If we perform the inverse operation, multiplying these covariant components by the contravariant metric tensor, we recover the original contravariant components, i.e., we have

$$x^v = g^{uv}x_u$$

Hence we can convert from the contravariant to the covariant versions of a given vector simply by multiplying by the covariant metric tensor, and we can convert back simply by multiplying by the inverse of the metric tensor. These operations are called "raising and lowering of indices", because they convert x from a superscripted to a subscripted variable, or vice versa. In this way we can also create mixed tensors, i.e., tensors that are contravariant in some of their indices and covariant in others.

It's worth noting that, since $x_u = g_{uv}\, x^u$, we have

$$s^2 = x_u x^u$$

Many other useful relations can be expressed in this way. For example, the angle θ between two vectors a and b is given by

$$\cos(\theta) = g_{uv}a^u b^v$$

These techniques immediately generalize to any number of dimensions, and to tensors with any number of indices, including "mixed tensors" that have some contravariant and some covariant indices. In addition, we need not restrict ourselves to flat spaces or coordinate systems whose metrics are constant (as in the above examples). Of course, if the metric is variable then we can no longer express finite interval lengths in terms of finite component differences. However, the above distance formulas still apply, provided we express them in differential form, i.e., the incremental distance ds along a path is related to the incremental components dx^j according to

$$(ds)^2 = g_{uv} dx^u dx^v$$

so we need to integrate this over a given path to determine the length of the path. These are exactly the formulas used in 4-dimensional spacetime to determine the spatial and temporal "distances" between events in general relativity.

For any given index we could generalize the idea of contravariance and covariance to include mixtures of these two qualities in a single index. This is not ordinarily done, but it is possible. Recall that the contravariant components are measured parallel to the coordinate axes, and the covariant components are measured normal to all the other axes. These are the two extreme cases, but we could define components with respect to directions that make a fixed angle relative to the coordinate axes and normals. The transformation rule for such representations is more complicated, but each component can be resolved into sub-components that are either purely contravariant or purely covariant, so these two extreme cases suffice to express all transformation characteristics of tensors.